PUBLIC POLICY DEVELOPMENT

PUBLIC POLICY DEVELOPMENT
Linking the Technical and Political Processes

ROBERT F. BAKER
RICHARD M. MICHAELS
EVERETT S. PRESTON

A WILEY-INTERSCIENCE PUBLICATION

JOHN WILEY & SONS
New York • London • Sydney • Toronto

Library of Congress Cataloging in Publication Data:

Baker, Robert Fulton, 1917–
Public policy development.

"A Wiley-Interscience publication."
Includes bibliographies and index.
1. Technology—Social aspects—United States.
2. Technology and state—United States. 3. Decision-making. I. Michaels, Richard M., joint author.
II. Preston, Everett S., 1911– joint author.
III. Title.

T14.5.B3 309.1'73 74-32157
ISBN 0-471-04435-0

Printed in the United States of America

10 9 8 7 6 5 4 3 2 1

PREFACE

Are objective analyses usually available for making public policy decisions? Or are such decisions largely intuitive, their analytic justifications being prepared later? How do scientists and engineers analyze, debate, or even communicate about issues that require an allocation of public resources? Who influences whom? And who should do so?

These are the kind of questions that prompted the writing of this book. According to Buckminster Fuller:*

> Throughout the past state
> Of innate ignorance of the many,
> The informed few
> Told the uninformed many
> What to do
> So that the many's coordinate efforts
> Could produce most effectively
> The objectives of the few.

We feel that this need not be the case. Furthermore, decision making is present, whether it is the "many" or the "informed few" who reach the decisions.

Theories on public policymaking are available, particularly in economics. However, unless one assumes that monetary values can adequately represent all other values, economic theory does not provide the answer in other areas.

The overriding thrust of this book, therefore, is to identify as clearly as possible—and to call it policy development—the type of analyses that are essential to the making of good policy decisions. By whatever name, these predecision activities will become more critical as the world population increases and the complexities of a culture multiply.

The book is the product of our involvement for many years in the public

* From "Ethics," by R. Buckminster Fuller, *Saturday Review of Literature*, November 6, 1973, p.36.

policy process at the federal and state government levels. Any scientist or engineer who has worked within the context of government is inevitably struck by the practicalities of the management of public enterprises. On the one hand, the working politician comes to highly technological problems peculiarly ill-equipped to relate science and technology to specific social issues. On the other hand, the technologist in government is usually insensitive to the social and political considerations that underlie decision making. There has been, therefore, a significant and frequent failure by the technician and policymaker to communicate and function jointly in the larger social interest.

In an earlier age, this communication lapse was an important but not crucial gap. Today, however, in what we perceive as a period of rapidly expanding humanistic social values, the disparity can be a real threat to the future of Western social systems. This is true simply because so much of Western society is built around highly complex technology. If these skills are not coherently linked with the society's goals, the society loses control of its own future. And in the final analysis the American society must obtain control of technology for its human ends through the political process.

Hence we define policymaking as the marshaling of the intellectual and physical resources and their direction to satisfying the needs of the society within the constraints that it may choose to apply. Since so much of the policy process requires technologic proficiency, the linkage of science and technology with policy is critical.

This book is concerned largely with the development of a public policy process. It is based on the premise that there are rational procedures understandable by and useful to the policymaker. These procedures are essentially systems analytic, except that they explicitly recognize social and political dimensions in addition to the physical and economic ones recognized in the military and space fields.

No attempt is made to develop a truly closed analytical process. The emphasis is on the logic of a policy development and analysis procedure and on one of several alternatives within that logic by which the fuller implications of decision making in the public sector can be made. This process is meant to provide not a social decision but sufficient information, in a form that the political process can understand, about the direct and indirect effects of social choice. The procedures developed are intended neither to be closed nor to provide exact answers. Rather, they add a dimensionality and a range of measurable consequences of any policy action.

One reason for such an approach is the great uncertainty that inheres in most public policy processes; exact solutions, and methods to produce such solutions, are therefore largely inapplicable or the assumptions are so restrictive that usually they cannot be even remotely approximated.

Second, social and political dimensions in the public policy area are of such significance that they must be included in policy evaluation. Given the present state of the art in social measurement, however, they cannot be correctly combined with physical and economic models or measures. Hence analytic elegance must be sacrificed for simplified measures of dimensions that are of critical concern to policymakers.

The kind of analysis suggested is done before a program is initiated and is used to assist the decision maker in selecting alternative means to satisfy a social goal. Therefore, the procedure is designed to avoid prejudging either the solution or the decision maker, and so no weighting schemes are needed. If any weighting is to be done, the policy maker should do it, not the policy analyst. The decision maker should weight the direct and indirect criteria according to his perception of the social context in which the program will be applied. However, with an explicit policy analysis procedure, such weighting would and should become obvious and direct.

The book both implicitly and explicitly defines a role for the technologist. In the policy development process he must (1) provide an objective and complete analysis of the means to social ends and (2) communicate the results of those analyses to the policymaker unbiased. We do not use technologist in the restricted sense of the physical and engineering sciences. Policy analysis is multidisciplinary, requiring the analytical skills and tools of the social and behavioral sciences as well as the physical. Otherwise some of the most important techniques and the evaluations of direct and indirect effects of policy alternatives will be overlooked, leading to an inadequate study.

Scientists and engineers also have the same obligation and opportunity as other members of the society to act in behalf of their own perception of social needs and goals. They have the right to propound and to lobby for those policy alternatives that they think best meet society's needs. They can do so as individuals or as members of an organized group, as part of an aroused citizenry or by seeking elective office and thus becoming part of the representative group of policymakers.

This book reflects some of our philosophical biases. In our view, it is neither possible nor necessarily desirable to avoid such biases. To the extent that they can be identified, they are stated directly in the first chapters. One of these is that the social importance of technological innovation has been vastly underrated in Western history. It is, of course, reasonable to ask whether cultural development and change stimulate technological innovation or whether such changes are caused by the appearance of the innovation. In fact, both forces are at work and have a continuing impact on each other. However, recent research on the history of science and technology produces considerable evidence that technology has had more to do with social and political change in the Western world than is commonly recognized.

Technology pervades modern Western civilization. Our culture is utterly dependent on technology—physically, socially, and politically. Technology and its process, we think, are so intertwined with social process that it is essential that the two become far better integrated. Such an integration is the only way to obtain a social process that uses technology for its own larger human ends. Otherwise the social process will continue to be driven by technology, as we believe it has been for a very long while.

Given the structure of modern science and technology, this social abdication need not continue. If the society defines its own goals and purposes, science and technology can create the tools for obtaining those goals. A social requirement and a social investment are the essential conditions. The issue is how the society shall explicate and assign its priorities.

For us, the answer lies in the political process. If one looks at the fundamental importance of that process, the concern is less with conventional notions of power and control, conflict resolution, and conventional management. Rather, the concern is first with *policy,* or the identification of the needs and goals of the society, and the setting of the priorities among them. Hence policy development is the process by which goals are defined, and means to attain those goals are identified and evaluated. This book is essentially devoted to an explication of that part of this policy process.

The book is written for undergraduate and beginning graduate students interested in the relation between engineering and public policy and decision making. To the practitioner, this level should not be too simplistic and should be appropriate for developing an understanding of how to relate ongoing technical activities with social, economic, environmental, and political concerns. Thus an introduction is provided to some of the basic problems relating technology and public policy. It also provides an approach to increasing the usefulness of engineering and other technology to decision making. The objective of the book is to give practitioners and students a frame of reference and an approach to understanding some of the broader issues that must be considered in the application of technology to social and political processes. Hopefully, the book will stimulate and increase interest in the public policy process and lead to a degree of sensitivity and sophistication that is now sorely lacking.

There is, or could be, a great deal of similarity between policy development and other policy analyses such as policy planning, strategic planning, management by objectives, PPBS, technology assessment, environmental impact, and social or economic impact. The term policy development is preferred because it is descriptive of a necessary component of making policy decisions and, therefore, differs from policy analyses for other purposes.

The book is organized into three parts. Part I provides a foundation and practical background for policy development. Part II consists of familiar examples. Part III serves as a basis for expanding and increasing the capability for conducting policy development studies. The relatively narrow set of examples represents the narrowness of our experience rather than that of the concept.

We acknowledge the assistance of many persons whose encouragement and comments have been very helpful.

The Task Committee on Research on Policy Development of the American Society of Civil Engineers, chaired by Dr. Roscoe F. Ward, University of Massachusetts, aided in clarifying terms and basic concepts. Mr. Walter G. Heintzleman, Port Authority of Allegheny County, and Dr. William S. Butcher, University of Texas—both members of this committee—made significant contributions to the discussions of the real world of political issues.

Two reviewers who deserve special thanks are Dr. Russell C. Jones, Department of Civil Engineering, the Ohio State University, and Dr. Duane W. Hill, Department of Political Science, Colorado State University.

A special mention is also due to the contribution of Dr. Thomas H. Dudgeon, whose editorial skills provided the coherence necessary to achieve the goals of the book.

Robert F. Baker
Richard M. Michaels
Everett S. Preston

December 1974
Bethesda, Maryland
Chicago, Illinois
Columbus, Ohio

CONTENTS

PART 1

THE POLICY DEVELOPMENT CHALLENGE

There is around us the metamorphosis of a society-that-has-been into a fresh, new, robust society-aborning. Engineers, scientists, and technologists provided impressive tools to cities, and produced mobility and machinery of yesterday's society. In the emergent America and throughout the world, those skills are equally essential, but they must be employed in new ways.

Our society is outgrowing a feudal relationship between the government and its citizens. In spite of the lethargy of both, we are emerging into a new culture where government and citizenry have indeed found a common cause: the freeing of the individual from domination by environmental (both natural and man-made) and economic forces beyond his control.

The emerging question that is addressed at every turn is not, as in the past, "Can we do it?" but rather, "Ought we do it?" Too many highways have impersonally run through parks, neighborhoods, or scenic areas; too much aircraft noise has flooded the quiet places; too many ticky-tacky, high-rise, concrete file-cabinets have been lined up as urban shelter where communities are needed; too many big, sprawling school buildings and campuses have been obstacles in the path of an individual's "hunger to know" from a teacher not an institution.

In general, and for a myriad of reasons, a substantial and/or vocal part of the American public has come to insist on having a public say in how technology is to be utilized and where or whether even perfectly engineered works should be built. We have arrived at that point where we need to know the consequences of our public actions—particularly resource use—*before* we take the action. This can be done only if a disciplined and objective effort is made to define our social goals and analyze the policy alternatives necessary to achieve those goals. The thesis of Part I of this book is that a viable process of policy development is known, and that it can and should be used to further the interest and development of the society.

Such a process requires that scientific, technological, and political skills be combined. Without a sophisticated integration, the chances of meeting social goals of a complex society are dim at best.

Furthermore, it is asserted that scientists, engineers, and technologists representing all fields of organized knowledge have a specific challenge to participate in the public policy development process. Only by so doing can the allocation of finite resources be done so as to support the technological future most compatible with the social goals. Thus the discussion is one of matchmaking, in which it is asserted that those who populate the wide world of technology must necessarily become an inherent part of the political process.

CHAPTER 1

SCIENCE, TECHNOLOGY, AND PUBLIC POLICY

Until well into the twentieth century, man was in constant conflict with his physical environment. His energies were devoted mainly to wresting control of his safety and security from natural forces. Famine, flood, pestilence, and plague have been a central and overriding concern of all emerging societies. Indeed, survival societies are characterized by the means they used to obtain life-sustaining elements from a very hostile and unpredictable environment. The very nature of the rules of social life in such cultures derived from the natural balance in the environment where they were located.

Regardless of the level of human development, it has been the eon-old struggle of man with his physical environment that has dictated and directed his energies. How to gain control over and provide a stable environment has been one of his dominant concerns. Man has used only two basic means to deal with his physical predicament. One has been religion acting through social control, the other has been technology. Religion was useful as a means of inducing conformity among members of a social group, and thus concentrating and sustaining human energy in available ways to mitigate the survival problem. It did not, however, provide the means to control and improve his physical environment. This was, as it still is, the province of engineering and technology. In this chapter, the changing role of technology in social organization is examined, and the emerging needs for a more effective process for policy formulation are discussed.

TECHNOLOGY AND SOCIAL ORGANIZATION

It is easy to trace the development of human societies to the emergence of some technology. In the context used here, *technology means any physical or operational methods developed by man that increase the efficiency of a human operation.* The ages of human development can be characterized by the technological innovations that changed the level of productivity or the amount of environmental control the human society could exert. The Stone Age, the Bronze Age, and the Iron Age are landmarks because each of them represented an innovation in the natural process which markedly changed the character of human society. These innovations were important because they increased the extent of man's control over natural forces, and allowed him to employ everincreasing amounts of his physical and psychological energy toward social and cultural development.

Technological innovations have been more than merely amplifiers or conservers of human energy. They appear also to have been fundamental determinants of social organization. The invention and application of the stirrup by the Huns gave them mobility and control over the horse that permitted them to overrun the Roman Empire. The result was not merely an armed "victory"; it was a transformation of Western Europe into a feudal society. Similarly, the invention of the steel plowshare revolutionized German society. A major revision of property lines followed its application which changed the power relations within the society. One result was a radical change in political organization.

Technological innovation and natural disaster have been two basic although certainly not the only shapers of social organization. Yet such innovation in human societies appears to have occurred as both a rare and almost random event throughout most of history. It has come by invention, generally by an individual who obtained an insight into a process and then "tinkered" it into existence. Only after it became functional and the inventor was able to demonstrate his own increased powers using it, did it spread its influence. This kind of random process is a very inefficient one and is probably one of the reasons for the very slow development of social organization. Another, of course, is that when men are trained to tradition, they are less likely to seek new methods. Life is closed and complete for such men; they see no conceivable alternatives to the traditional methods. Innovation requires a different outlook on nature, either because of individual idiosyncrasy or a failure of the tradition to meet the needs. Innovation or ingenuity inheres to the root of the words "engine" and "engineer."

Organization of the engineering, the innovation process, did not occur until well after the Renaissance. It probably emerged out of the craft system, whose closed groups of skilled artisans passed on methods, tools, and techniques through apprentices, but always within a closed social system. These craftsmen, along with the small businessmen, formed the middle and upper classes out of which the modern age was born. As industrialization developed, these groups became the core of production managers and the creators of engineering specialities.

The crafts, as well as the early engineering skills, were based primarily on lore and experience, and came out of the guild halls through the apprentice system. Engineering as a "profession" emerged from this process and based its uniqueness on empirical grounds. There was no formal body of theory, only past trial-and-error experience that sorted out what had worked and what had not. Its principles were a series of dogma; for example, "foundations shall be carried to solid bottom." (It was another 200 years before the scientific principles underlying engineering design were defined and the process changed to an analytic one rather than an empiric one. And it was the oldest of the engineering specialties that changed last: civil and mechanical.)

The central nature of technological innovation in social development converged in the eighteenth century with the emergence of engineering as a formalized craft. The consequential industrial revolution began in the eighteenth century and was destined to be one of the significant forces leading to the overthrow of every social system in Europe, the overturn of centuries of political philosophy, and to the creation of the foundation for a humanistic world order that would emerge in the last half of the twentieth century.

To a significant degree, the engineer was responsible for the creation of the industrial revolution. Industrial processes were the product of mechanical engineers, who harnessed first power, then steam, to operate vast assemblies of production machinery. Mechanical devices were such enormous amplifiers of human and animal power that it was possible to increase productivity by several orders of magnitude. The emergence of production machinery and volume production immediately increased the necessity for transportation facilities. Increased capacity was essential to bring coal to the plants as a source of energy and to carry the production to market. Civil engineering joined with mechanical to develop barge and canal systems and ultimately the railroad. In essence, the engineering skills were applied to provide the technological infrastructure for production and distribution. Only after such engineering systems were created could modern economic philosophy be created. An Adam Smith or a Karl Marx could not have existed in feudal social order. They could be products only of a technological era.

Some Consequences of the Industrial Revolution

The advances in productivity and in control of the physical environment were not the only consequences of the emergence of engineering and the industrial revolution. Major social effects of this technological process included the change in structure, roles, and values of all societies that it touched. Man's perception of his place and security within his society was determined to a great extent by the industrial process. What he could do and how he could satisfy his own needs were predefined by the industrial skills he could acquire and the competitive role that the industrial system permitted him.

The industrial and mechanical process systems determine uniquely, by their design, the kinds of skills that are required, the native talents that will be effective in operating the systems, and the physiological criteria for effective performance. More important, the whole reward structure within a society becomes based on the capacity to perform within the context of industrial processes. Indeed, whole pay and promotion systems have been developed under the rubric of job classification. In essence, engineering design determines a man's capacity for security, rank, and prestige within a community. Even though the personal risk and the energy requirements of the coal miner are much greater than those of the machine tool operator, it is the latter who receives higher pay and higher status in the industrial society. This is an almost complete reversal of values from the preindustrial age.

More striking is the fact that this industrial value structure is based on man-made systems bearing only the remotest connection to traditional social criteria of worth. Wealth, measured in terms of chattel and/or money, became the dominant goal of men and their measure of worth, rather than agriculturally productive land, which was the measure of wealth and power in feudal society. From a political standpoint, it was trade and the resources it provided the state which made industrialization attractive.

Among the consequences of this process was the destruction of the aristocracy in Europe which had used land to provide wealth and power. Loyalties were bought and sold with land. In an industrual age, the relative value of land diminished and wealth and power were being measured in different ways. It has been argued by several historians that the American

Civil War arose from a conflict between an emergent mercantile northern society, and a feudal agrarian southern culture. *In sum, the industrial revolution was a basic vehicle for ending forms of government that had existed with only differences in scale from the beginning of recorded history.*

Along with a change in values, reward structure, and political form, the industrial revolution changed social geography. To support major industrial development, three conditions had to be present: energy, labor, and transportation. These dictated the location of industrial complexes in coastal regions or on major waterways. It also required a shift from rural to urban population distrbution. Only in this way could the concentration of labor to operate the industrial complex be obtained. Using all forms of inducements, as well as a variety of less savory practices, a large proportion of rural agricultural population was slowly urbanized.

The handiwork of the engineer provided the primary means for transforming the feudal agrarian world into a form categorically different from anything the world had ever known before. Man virtually wrested control of his destiny from the tyranny of nature. He learned to modify natural forces to serve his own idiosyncratic ends. He learned to reorganize the physical environment to produce levels of control never before achieved and to reduce the variability of natural processes to within tolerable limits. By the engineering of water supply networks and waste collection and disposal systems for cities, it became possible to manage high density concentrations of people and provide significant improvements in physical security for large masses. Thus the industrial city was born. In the end, man overthrew the tyranny of nature and substituted the tyranny of technology.

There should be no doubt that the industrial revolution was a tyranny. In the eighteenth and nineteenth centuries, and well into the twentieth, this tyranny was as despotic as any known in human history. The mechanical systems which engineers designed required special skills and special emotional requirements from the labor force. The process of selection was ruthless, dictated by the machinery. Those unable to do the work because of emotional, physical, or intellectual limitation were left to starve. To survive, many families had to indenture every member to the industrial process. Sickness, which in the agricultural life was readily managed by the extended family, in the industrial environment was a disaster, for it meant instant loss of financial security.

It is not unreasonable or unusual to identify the industrial revolution as a technological tyranny. It is curious, however, to suggest that the engineer has been the essential tyrant. We are used to thinking in terms of Western mythology that the tyrants of the industrial age were the new aristocracy: capitalists, the moguls, the magnates, and the robber barons. The Western myths have frequently confused power with control. The notion of aristocracy, regardless of form, has always been a symbol of security, and hence endowed with awe. Yet the real control lay, as it still does, in the hands of the engineer who designs, guides, and directs the total production process. It is his performance which ultimately determines what can be produced, how efficiently, and therefore at what price. It is true that the corporate managers may refuse to adopt innovations, but they cannot create them. It is equally true that labor can refuse to operate the tools of production, but they cannot create them either. In essence, both the aristocracy and the laborers have been indentured to the engineer.

To a large extent, the engineer has been a servile tyrant. This may well be due to social location of the engineer in Western society. His was a craft; he was an artisan. Thus the engineer evolved, as a social group, into the middle class of society. Its membership has, and still does, come from the middle and lower middle classes. As a consequence, engineers have generally been satisfied with middle positions in their social institutions and have been satisfied with rewards of a middling sort; middle income and monuments engineers could identify as their own, such as bridges, highways, ballistic missiles, or sewage treatment plants. The engineer has not recognized his impact on or the extent of his actual control over Western development. He has controlled passively out of the center of every society in which he has existed. He has eschewed the political and social choices, but has aggressively sought to make the technical ones.

Convergence of Science, Technology, and Engineering

Two emergent trends in the Western world that led to major changes in the technological process were the growth of science and group innovation. These emergent trends have led to major changes in the technological process. Both have converged in the last half of the twentieth century. The one we are concerned with is the growth of science. From a social and hence policy standpoint, the evolution has changed both the nature and power of the technological process.

Science is a thought process. It is a form of logic for deriving conceptual relations about physical processes. Science is concerned with the manipulation of concepts while engineering is concerned with the manipulation of objects. This distinction although an oversimplification is important because it placed science and engineering on two independent, if parallel, paths. The linkage of the two was negligible until the twentieth century. After all, science was for much of its history a part of philosophy. It was far too abstract for the early engineer who was concerned with the twin questions of "what?" and "how?" The relevance of the question "why?" was secondary—a philosophical and academic question. The engineer's was an empirical tradition, not a theoretical one. The latter, indeed, was the scientist's concern—interesting, but of little practical utility.

The path the scientist took has mainly been to develop explanatory or descriptive principles. Science produced "laws" like Ohm's or Hook's which are largely descriptive of the relations among physical variables. However, unlike most engineering rules, these descriptions had a significant property: they related variables in a functional way which permitted predictions of effects caused by the changes in those variables. Experimental science evolved as a means of testing the validity of the predictions in an observable context. This laboratory process produced broader generalizations about the nature of physical properties than the strictly empirical methods. Thus general principles of even broader generality and which were increasingly consistent with real world observations were constantly sought. Functional relations were the hallmark of science, regardless of the domain of interest.

By the end of the nineteenth century, the thought process called science was providing something unknown before in human history. First, there was an ever more coherent understanding of the interrelation among physical processes. Second, and more important, there were means of making the natural processes more fully predictable. Science abstracted from physical experience a perfectly general meaning and communicated that meaning with the economy of a mathematical equation.

Science is, of course, an art form, but one that has had a singular power: to change our conception of the world of which we are a part. The other arts are by nature emotionally evocative where science structures our perceptions. In this sense, it is an amplifier of man's powers of relating to the physical world. It has, consequently, shattered almost all traditional myths about man, his universe, and his relations to it. By using a different semantics—operational logic—science provided the means to escape from philosophical abstractions and made the functional connections among real events and perceptual experience.

The ultimate power of the scientific method inevitably converged on traditional engineering. The modern basis of structural design is founded on Newtonian mechanics. Aerodynamics is based on fluid mechanics. Engine design is based on thermodynamics.

Thus the convergence of science and engineering led to a change from empirical design to analytic design. The development of mathematical models of increasing complexity provided a body of design theory which converted many aspects of engineering into routines. The models also provided a means to test wholly innovative designs as well as to evaluate design concepts, no matter how they may have arisen. The scientific work of men like Guilleman, Shannon, Gibbs, Weiner, and others have provided basic principles for understanding the processes by which entire systems could be invented. Thus where before, innovation was the accident of individual insight, today's scientific theories permit innovation by design. This is accepted as commonplace in most engineering disciplines today.

Nonetheless it is easy to underestimate the magnitude of the change in the method of system design which the body of scientific theory has brought to man's physical creativity. For example, in many of the more advanced fields, the empirical testing of design concepts has been eliminated. The understanding of the design process is sufficiently complete that the outcome of any design is perfectly predictable. Uncertainties, estimates, and the effects of variability are eliminated by the established theories to shift the weight of concern from construction to innovation. Thus the emergence of scientific theory across the spectrum of physical sciences has revolutionized the engineering process in the twentieth century. New developments and systems are designed rationally and depend on the availability of the appropriate science. The transistor is the transformation of a theory of solid-state physics into an operational device. Nuclear fission or fusion was not conceivable without the nuclear physical theory developed in the first quarter of the twentieth century. In sum, scientific theory has become the directing force behind technological innovation. So intimately have the two things become intertwined that we now speak of science and technology as a single concept.

As a consequence, a whole series of science-based fields have emerged in the past 30 years. The result is that where originally the practicing engineer saw little utility in science, modern science and technology now sees only limited relations with what has been known as engineering. In high technology industries, traditional engineering is now production processing.

Another marked change in the nature and productivity of science and technology was its institutionalization during the middle of the twentieth century. As noted earlier throughout all but recent history, innovation was an independent, almost random event. Early in the industrial revolution, it became clear that technological invention was a valuable thing. It became increasingly desirable to induce invention and reward it. Since those innovations of highest importance were mechanical devices and systems, a patent system was conceived and created very early in the industrial revolution. It was, of course, an incentive system that provided inventors with the hallmark of the solitary creator and the opportunity to reap personal economic rewards for creativity. Clearly, this bespoke the recognition by the aristocracy that the innovations needed to increase productivity and create capital came from the inventor, who was often a member of the peasant-rooted engineering guilds.

The patent system worked remarkably well as an incentive system early in the industrial age. Among engineers, the number of patents a man held became a measure of his success in an achievement-oriented economic order. Patents were to the engineer's status in his social group what publications were to become for the scientist in his subculture. One of the results of the patent system was a marked increase in competition for technological innovation. In its early years, this was a highly individualistic competition.

A major change began to occur around the end of the nineteenth century. Industries had been formed and consolidated around major technologies. Mining, metals, textiles, and machinery production systems had emerged in modern form. Differentiated corporate entities had been created. Among their divisions was always an engineering one. These engineering groups not only directed the production process, but also were charged with the improvement of those processes. The coalescence of like specialties in a single group of achievement-oriented men sharpened competition among them and increased the rate of product improvement. As elsewhere, the establishment of homogeneous groups induced conformity and gave institutional goals added force. The engineering value system was based on inventiveness, achievement, conformity, and productivity. Bringing these groups into being also increased technical dialogue, and this communication amplified technological creativity both in range and sophistication. Finally, within the industries, these engineering subgroups shared the major values and goals of the corporation.

Group Innovation

One consequence of these subgroups was that the corporate—and ultimately government—engineer assigned his patents to his company. He retained the title in his name (which always appeared on the patent), but the economic reward for the device was transferred to the corporation. The reward for the engineer became status and promotion within the company. His reward frequently bore little relation to the economic gain the invention might produce for the corporation. The security he derived from this relation, and his close identification with the values of the company, were sufficient to make this trade-off acceptable. The company engineering departments thus coalesced around inventiveness in behalf of and loyalty to the corporations or government agency of which they were a part. The marked increase in patents which this process caused was notable. Among other things, it changed the patent process from an incentive system to a protective mechanism for corporations. This, in turn, has brought the patent system to a point of little importance in technological innovation.

Long-term protection of rights does not matter any more. The high rate of technological change that institutionalized research and development produces leads to ever-shortening product life. Hence it is the capacity to maintain innovative leadership that determines share-of-market for the corporation. For this reason, most modern corporations determine their investment in research, development, and engineering in the terms of projected rate of return in the marketplace. The relation between patent life and product life, therefore, has all but disappeared. Marketing has become a far more productive way of being competitive than use of the patent system.

FORMULATION OF PUBLIC POLICY

The United States, as well as other parts of the world, was in the process of change in the midtwentieth century from a value system derived from the security or safety needs of history to one based on the quality-of-life needs or desires of members of the community.[1] The new emphases on concerns were about such things as environment, esthetics, discrimination, disadvantage, crime, and health. These are all elements of human life style that bound the quality of human existence as contrasted to mere survival. The nature of public policy-making within this new value set is the major focus of this book.

Public policy becomes more central and significant in an age whose values are centered on quality of life, rather than on simply material well being and physical security. When survival was at stake, policy was concerned with controlling and directing human behavior to ensure subsistence. When security was of primary importance, policy was concerned with the development of the tools of production, the refinements of social interchange, and the resolution of social conflict. In sum, the purpose of public policy has been one of establishing the means and managing the resources to enhance social development and order. In the age of the new values, these purposes are no longer sufficient. They are such a narrow set of concerns, so short-term in their focus and so biased toward special interests, that there has emerged disconnection between governmental actions, and individual and informal group expectancies. The history of the decade of protest of the 1960s with its demand for participatory democracy is evidence for the incompleteness of the policy process.

Public policy is a mechanism for realizing social goals. The crux of the process is that it provides a valid reflection of the goals and priorities of the

[1] The transitions from a survival to a security to a community culture are discussed in more detail in Chapter 7.

citizenry. In the transition to a community age, this is a fundamental political problem within the American society. It is reasonable to hypothesize that the clear-cut lack of trust of political institutions in particular is the consequence of the disparity between political goals and the priorities of the individuals within the society at large.

The reason for dissatisfaction becomes clearer when one looks how public policy is developed. Public policy development in the end involves three operational activities: goal setting, resource allocation, and evaluation of effectiveness.

Goal setting is essentially the identification and specification of the priority needs of the society. In democratic forms of government, the elected are charged with representing and reflecting the values, needs, and goals of the electorate. The theory is that the elected are the connecting link between the society in the disaggregate, and the governments becomes either a tyranny or a special-interest protector. Tyranny in this sense means an institution exercising power to achieve goals that are not consistent with those of the society as a whole. Such disparities may arise not only when the elected are nonrepresentative, but also when responsibility for policy decisions is delegated to administrators in the executive branch. Bureaucrats in the governmental process are measured according to their record of executing, of carrying out, of doing. Therefore, unless the ends which their means allow them to reach are directed by the political process, they will implement their own goals. Events of the 1950 to 1970 decades in the United States suggest a steady trend of the executive branch to usurp the first and fundamental responsibility for the elected official politician—setting social goals.

The second function of public policy is resource allocation. Given a set of goals in some priority order, it is then necessary to decide how the goals may be attained. The issue is one of evaluating alternatives, determining the requirements to create those alternatives, quantifying the physical and human resources required to implement the chosen means, and establishing the time requirements to reach the goal. The evaluation of alternatives is crucial, for it is the only way in which policy goals may be transformed into operational form so that the ends may be achieved.

There is an intimate relation between goal setting and resource allocation. Unless goals are made explicit there is no way to allocate resources rationally, nor is there any way to predict whether any allocation decision will achieve the desired goal. Conversely, no matter how well defined a policy goal may be stated, unless the resources allocated to the chosen means are sufficient, goal attainment is unlikely. Clearly, methods of analysis, specialized knowledge, and technical skills are needed to evaluate all means of attaining policy goals and to establish the appropriate levels of resource requirements. Technical skills are tools by which the decision-maker may obtain necessary information in appropriate forms.

The third function of public policy development is evaluation of effectiveness. Both goals and means to achieve goals have considerable inherent uncertainty. Obviously, any program selected as the means for achieving goals is not certain to be successful. Furthermore it is rarely possible to predict in advance all the direct and indirect effects of a selected program. Thus only by a continuing process of rational evaluation can the effects of a public program-in-being be determined. Without this process there is no way to determine whether such a program should be continued, modified, or discontinued. The evaluation of public policy is one of the least recognized and the least effective phases of the public policy process.

A major consequence of an inadequate evaluation of public policy is a proliferation of programs. Without evaluation, a program can be funded without measurable evidence of effectiveness or productivity relative to reaching the desired goal. As a result, an increasing number of programs are not cost-effective. But worse, as new programs are added to old, the resources expended on each of them grow rapidly. Without a continuing and comprehensive evaluation process to guide resource allocations, the costs of government increase as the returns to the society decrease. Also, because resources are always limited, the capacity to meet new needs is diminished and emergent requirements are deferred or are unmet. Finally, failure to evaluate programs continually will diminish innovation and the utilization of new methods and materials to achieve desired goals. This will occur simply because without a complete policy development process, failures go untreated until they reach crisis proportions. For example, if a country contains 6 per cent of the world's population and produces 15 per cent of the world's food, can there be a satisfactory justification for any of its citizens' going hungry? Since agriculture is regulated through public policy, a failure in the distribution of food constitutes an inadequacy of the policy process itself.

The Needs of Public Policy Development

One can talk both abstractly and specifically about the needs of public policy development. In the following discussion, both are considered under conditions that existed in the United States in the early 1970s.

How can public policy-making be responsive to new values and a different class of social needs? The answer to this question is not fully clear, but boundaries of the emergent social needs can be identified. There are at least four classes of values involved. First is a fundamental concern for the primacy of the individual, and second is his opportunity for personal development. These values are based on the premise that society has an affirmative responsibility to assist and support the individual in his growth and development. A third class consists of the qualitative and esthetic aspects of the social and physical environment. This includes not only the support for the structure and form of art but also the nature and level of social dialogue. The fourth is conservation of the social and physical environment and the enhancement of the connections between those environments and the individual.

This type of boundary condition provides the minimal social parameters within which to evaluate public policy. Effective public policy making can and should be measured against its social responsiveness in these domains.

This framework leads directly to the question of what tools can the policy-maker use to define goals and to modify social structure to achieve those goals. To answer these questions, one can start with the domains within which public policy operates. In the United States, public policy operates in four major substantive areas: public works, social welfare, economic control, and knowledge production.

Government also performs any necessary function of regulation and management of resource allocation. In the United States these functions are quite limited. In part, they are diluted by the magnitude of the private decisions and resource allocations within the society. Government carries on transactions with these private activities and adjudicates conflicts, acting in the public interest. However, government does not exercise control over the resource allocation and distribution conducted by private operators. The international energy crisis was illustrative of this point.

The American social system in the early 1970s was a pluralistic society in which private decision-making overwhelmed government decision-making in most areas. Public policy-making was thus inherently limited to those systems which fell outside the influence of profit-making forces, using private enterprise to activate the system. In this sense, the public policy-maker had relatively few societal systems that he could manipulate to achieve larger social goals.

Often a concept or a social innovation is feasible that would serve to meet or at least provide an increment to attain the desirable social goals. However, they may be impossible to implement because they generate strong social resistance. A classic example is in the area of land use control. There were those who contended that it was in the public interest to develop land for social purposes in a coherent and rational way. Yet the value of land to private enterpreneurs was so great that attempts to seriously control its use were rejected in the early 1970s; that is, it was not politically feasible in the United States to consider land as a public resource rather than a private commodity to be used for private profit.

But probably the key restrictions on public policy-making are the technical constraints. Even where a social goal is explicit, the decision on the means to achieve it has been severely constrained by technological limitations. For example, no one was really satisfied with the educational processes of the early 1970s, but there were few ways known that would produce a significant improvement in that process. (There is, of course, no reason to believe that even if a way were developed, that it would be socially acceptable.) In still other cases, there may be technologically feasible concepts for achieving a social goal, but the costs and time to develop and implement them are outside the realm of feasibility.

The ultimate consequence of these constraints is to restrict severely the domain of public policy. On the one hand, the policy-maker cannot impose politically or socially unacceptable changes on the society. On the other, he lacks the knowledge to identify or evaluate innovative solutions. The result is that policy-making is usually conservative in the sense that past practices, methods, and technology are the basis of policy-making. Not infrequently, therefore, policy-making is a prisoner of history, technological as well as organizational. Public works is a most obvious example, and public works policy is essentially determined by tradition, in this case engineering tradition. In an era of changing social values, dependence on such traditions is decreasingly acceptable. Yet the public policy-maker has no real alternative. He must waste his energies in resolving conflicts resulting from its application, rather than in selecting and using technologies that are designed to satisfy social goals without degrading the social and physical environment that should be preserved.

In sum, public policy-making is caught in a dilemma of major proportions. The process must define the goal or set of goals consistent with those of the society. The process then must find, develop, or select a means that will provide at least a step toward that goal and one that will not compromise other goals of equal importance. Yet the policy-maker has few program alternatives at his disposal that will allow him to accomplish this. In addition, he has few ways of evaluating any alternative presented to him. If he is provided more than a single alternative, he has no way of comparing the impacts on criteria important to him. Rarely is there an opportunity to recognize alternative indirect costs and benefits. Thus it is difficult, sometimes impossible, for a policy-maker to determine which of a set of alternatives will achieve a social goal most efficiently, most quickly, at the least direct cost. Yet rational policy-making must include these kinds of evaluations as well as the social and political ones.

Thus the public policy-maker is constantly in a state of means-ends ambivalence. If he knows the social goal that he wishes to attain, he does not know the most effective means to attain the goal. If he knows the means, then he does not know the goals which those means can achieve. In developed societies, and in the United States in particular, the latter part of this dilemma is the most troublesome to the policy-maker. New methods and new technology keep emerging and intruding. Pressure is exerted for their use. Almost always, some social benefit is included as justification for public support and/or investment. The real social goal which new methods can or will serve is never made explicit. This was true for consideration of highway transportation when individual mobility was not seen as an explicit goal.

The other side of the coin is the frequent confrontation situation of a group seeking use of a public policy program to satisfy the vested interest needs of the group. Thus, there were "poor" marches, trucker slow-downs, consumer boycotts, and so on. Normally, the public policy-maker has not had the knowledge or the capability for rational decision-making in these situations. This has not precluded decisions being made and resources being allocated, but there have been poor decisions and less than satisfactory allocations.

The American public policy system has historically worked on the bases of random forces. Hence the policy-maker has frequently been caught up in applying means without knowing what social goals will be served or whether they are in any rational priority order. Furthermore, the policy-maker rarely knows what the true social pay-off will be and often he does not have an adequate idea of what the true social, as well as economic, costs will be. This problem is very clear in relation to the space program in the United States. What was the real societal goal that this investment was to satisfy? What was its true social cost? That is, what other programs of social significance could those resources have been used to serve? None of these questions was explicitly asked or answered before the policy decision was made. In the face of a perceived Soviet challenge, and because it was technologically feasible, the investment decision was made. The space program is not an unusual example of American public policy-making. One can look at a whole range of public works programs and find the same process at work.

One major effect of this kind of public policy process has been that the policy-maker has been far more concerned with program management than evaluation of program effectiveness. For example, an overwhelming majority of states established motor vehicle inspection programs even though there was almost no evidence that such systems influenced the accident rates in those states. Certainly there was no evidence that these systems were cost effective. Yet they received tax support, and state legislatures as well as the federal government expended continued efforts to increase their scope, organization, and management efficiency.

In the 1970s, the American policy-making process was overwhelmingly concerned with system operations and not system performance; with budgets and not productivity per dollar of budget. Policy-making had become structurally oriented rather than performance oriented, management oriented rather than social-goal oriented. One result of this type of process is a major breakdown of the governing system, since it will not appear to have produced results of benefit to the society in the form of achieved social goals of priority importance to people. Such a process appears to be dominated by special interests rather than by social concern. To a significant degree, along with the use of past history as the basis for policy making, this led to the upheavals of the 1960 decade, the dissaffection with and mistrust of the political system.

Within the context of the American political system of the early 1970s, modifications of the policy process became essential. Policy-making and executive actions were separate as well as different. Since the primary function of policy-making is goal setting and resource allocation, including, of course, the evaluation of both, one of the basic reasons for separating powers under the constitution was to differentiate the two. It has been an interesting twist of the twentieth century that the election of a President has come to mean a mandate for leadership. The Federalist Papers are quite clear in describing the presidency as an executing office, over and opposed to the Congress, which was conceived as a policy-making body. The violation of this principle which emerged as a consistent pattern in the middle half of the twentieth century contributed to the continuation of policies and programs over which there was little rational control. This fact, coupled with the appointment of department heads responsive to constituencies within the area of departmental interest, has meant that the likelihood of innovative policy being developed was slim to none. Finally, the linking of legislators to specific executive programs so distorted the governmental process that little or no rational policy development or evaluation was possible.

Linking the Technical and Political Processes

Rational policy development includes a rational analysis for goal setting. Certainly as far as science, technology, and engineering are concerned, there is simply no way to design systems unless goals are clearly stated. Unless there are clear statements of societal goals and open debates on priorities, there is no real possibility of looking at or evaluating alternative means to achieving goals. Goal setting is the foundation for the whole policy process. Without it the society cannot focus its resources or its energy. It is also clear that social goals must be established in a form more concrete than mere philosophical abstractions: they must be operationally defined. The major purpose of the goal statement is to provide the basis for action and the criteria for evaluation of alternative action programs. Unfortunately, most legislative bodies in the 1970s were spending too little time on defining goals—which they were the most obligated and qualified to do—and far too much time on debating alternatives technologies (e.g., highway versus mass transport) for which they were seldom provided adequate or objective evaluations.

Rational policy development also requires a procedure for evaluating program alternatives which may be proposed to achieve a policy goal. Any policy development process implicitly assumes that there is a set of methods by which alternative means for achieving policy goals can be compared and evaluated. Unless this is possible, public decision-making becomes largely unpredictable. The problem is to find methods that are rational and explicit so that decision-making becomes an open process rather than an intuitive one based solely on opinion, personal judgment, and group pressure. Traditionally, the public policy-maker used past history as a basic guide as to what was possible and projected that history into the future. Past programs, traditional technologies, and historic institutional arrangements have been supported and continued on the basis of prior utility. This is convenient in the face of uncertainty, but it is degenerate whenever the basic values on which those programs were based are different from the social values of the present or the future. It is a basic thesis of this book that the United States is a society in transition and that new values have emerged and are still emerging which make that dependence of policy-making upon historic tradition an error of profound consequence.

The preceding chapter has discussed the unlinked, uncommunicative relationship between the technical and political processes that existed in the United States in the early 1970s. The traditional role of science, technology, and engineering has been criticized on the grounds that an affluent society will choose the innovations it wants, and will resist technological tyranny. On the other hand, the inadequacies of the political process have also been identified, and many of their past failings traced to the inadequacy of their dialogue with technologists. Inadequate does *not* imply a return to the tyranny of technology—but to the evolution of a dialogue that allows the political process to allocate resources in such a way that will result in achieving the priority goals of the community. And so the basic needs of public policy development have been identified, and the character of the process has been described as the analyses that precede decision-making. We can proceed now with some particulars.

SELECTED READINGS

1.1 Bernol, J. D., *Science in History,* London: Watts, 1965.

1.2 Bronowski, J., *Science and Human Values,* New York: Harper, 1965.

1.3 Davenport, W. H., and D. Rosenthan (Eds.), *Engineering: Its Role and Function in Human Society,* New York: Pergamon Press, 1967.

1.4 *Engineering for the Benefit of Mankind,* National Academy of Engineering, 1970.

1.5 Meadows, D. H., et al., *The Limits to Growth,* New York: Universe Books, 1972.

1.6 Palter, R. M. (Ed.), *Toward Modern Science,* New York: Dutton, 1969.

1.7 Price, D. J., *Little Science, Big Science,* New York: Columbia University Press, 1963.

1.8 Price, D. J., *Science Since Babylon,* New Haven: Yale University Press, 1961.

1.9 Snow, C. P., *Science and Government,* Cambridge, Mass: Harvard University Press, 1961.

1.10 Wells, H. G., *Anticipations,* New York: Harper, 1902.

CHAPTER 2

SOCIAL GOALS— THE FOUNTAINHEAD OF PUBLIC POLICY DEVELOPMENT

The purposes of policy development are (1) to assist in the definition of policy goals and (2) to evaluate the probable effectiveness of alternative means of achieving policy goals. The process involves transforming these goals or requirements into some measurable form and then evaluating alternative means for achieving those goals against a set of performance criteria. Ideally, a common performance measure would be used across all dimensions so that an overall as well as an individual measure of projected effectiveness could be obtained.

In this chapter, the principles of public policy development are examined and the relevant implications on the setting of social goals are discussed.

DEFINITIONS AND PRINCIPLES

Policy-making is a part of the political processes under which groups of individuals, communities, and nations are governed. Policies that are established become the "rules of the game," and are a part of the political strategy for achieving the purposes of the group. Just as individuals must pick and choose between alternative benefits and expenditures, so must policy-makers select the compromise that has the greatest political feasibility. Far from being a precise, technical activity, policy-making is an intuitive process.

Policy development, on the other hand, can be much more analytic and much less subjective, for it is the process that compares the consequences of alternative policies. Policy development does not provide a decision; rather it provides an objective basis for a decision.

The distinction between an analytic approach for policy development and an intuitive basis for the decision process is important. To be more than a political banner, a policy must be very specific as to goals, resources, and constraints, and rarely can precise policies be derived intuitively. Analyses are needed that provide sufficient reliability that a policy can be successfully implemented. These analyses cannot be constrained to purely intuitive methods without serious detriment to policy-making.

Efforts to combine policy analyses with policy decisions lead to either making policy decisions a part of the technological processes; or extending the political process into technology. Neither of these extensions leads to the evolvement of the most effective policies.

While a clean distinction should be recognized between the policy-making and the policy development processes, the two will need to be very closely allied. A high degree of mutual confidence must exist. Policy-makers will want assurances that the alternatives have been objectively considered and that assumptions and priorities have not been inadvertently or deliberately omitted or obscured. They will also want *results of analyses* rather than *decisions* from the policy development process.

Definitions

The term *policy* is closely related to the word *politics*, and has grown in usage partly because the latter word has "undesirable connotations" of "partisanship and corruption."[1] Policy science is concerned with the study of human relations. *Politics* is the action phase of these studies; that is,

[1] See Selected Reading 2.3.

politics is concerned with influencing human relations, and the *political process* refers to the methods used.

Policy has been defined as "a body of principle to guide actions." It has also been described as the "most important choices made either in organized or in private life."[2] Thus a choice must have been made whenever a policy is encountered; that is, a policy cannot be "derived" in a mathematical sense.

Political strategy, or the means employed by partisan politicians, frequently calls for nebulous and idealistic statements that are designated as "policies." For this purpose, the aforementioned are guidelines for actions, rather than rules or legal requirements. For example, a policy to have cleaner air, purer water, or safer highways is sound political strategy, since most individuals are in favor of such conditions. The lack of specific goals is also good political strategy, since those who would be threatened by a more rigorous policy have no basis to resist the nebulous goal.

To deal effectively with improvements in the social organization, policy statements need to be more precise than is necessary for a political strategy. If they are not, debates and delays will result because there will be no quantitative basis for translating the full intent of the policy. Therefore, to be effective a policy goal needs to include (1) the goals or objectives in quantitative terms, (2) the resources that are made available, and (3) the means by which the goal is met. The emphases of this chapter are on the goals and resources, while Chapter 3 discusses analyses of alternatives.

General Procedure for Policy Analyses

Policy analyses generate a measure of effectiveness of the set of possible alternatives for a policy goal. Obviously, where the alternatives involve the modification of an existing societal system, the present performance of the existing system could be evaluated in the same manner and compared to the effectiveness of the alternatives.

Given the cost (time, money, energy, etc.) effectiveness of the alternatives, it is then necessary to evaluate the feasibility of the implementation of the alternatives. This is a problem of evaluating the rate of return in social terms, the time requirement for implementing, and the sociopolitical acceptability of each of the alternatives. Each of these evaluations requires a different method of analysis, but all three of them are equally important in policy-making.

The results should provide the policy-maker with a full range of the

[2] See Selected Reading 2.3.

possible means for achieving a defined social goal and estimates of the feasibility of implementing the alternatives. The process is an open one in the sense that the decision-maker not only makes the final choice but also has the freedom to weigh the various performance criteria in ways that best represent existing social values. Thus the process should and must permit final choice to be socially optimal rather than technically or economically optimal.

The general steps in a policy analysis are shown in the flow diagram in Figure 2.1 (see also Figure 3.1.) Policy analyses flow from both a goal setting activity and from a set of socially derived performance criteria, both of which are sociopolitical rather than technical in nature. Both must be provided to the analyst by the society itself. These criteria are essentially the dimensions of social performance which are of value to the society. If the society does not provide these criteria, they are assumed by the administrators or technicians. Since specialists are not usually representative of the society or its values, they can be expected to provide performance criteria at considerable variance with those of the larger society. The consequence is generally either dissatisfaction within or destruction of the society. Thus it must be the society itself, acting through its policy-making entities, that defines the performance criteria to which specialists must be responsive.

After the social decision has been made as to what criteria of societal system performance shall be used, the technical policy analysis process can begin. To evaluate an alternative system or means for achieving a given social goal, some form of measure must be applied to those criteria. Without such a set of measures, it is impossible to determine whether any

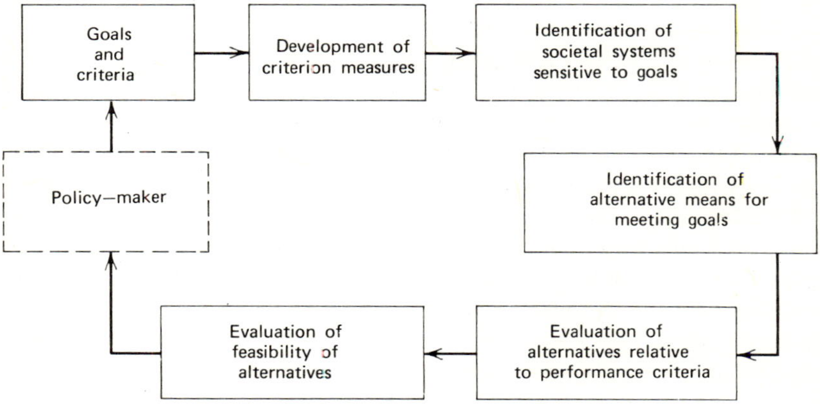

Figure 2.1 A general process for policy analysis.

policy will achieve its goal or whether any one policy will be better or more cost effective than another.

The next step in the process is to identify the basic social systems that can be modified or replaced as a probable means of improving the quality of social life. In many cases, if not most, attainment of a social goal requires an action on only one such system. Thus a mobility goal is most likely restricted to some aspect or another of the transport system. In many cases, however, several of the societal systems are involved if the attainment of a policy goal is to be obtained. For example, for certain goals relating to mobility, not only would transportation be included but so also would communication. It is essential then to identify all the societal systems that are means for achieving a given policy goal.

The next step is to identify an exhaustive set of alternative means that can attain the policy goal. Once a set of alternatives has been selected, each alternative is evaluated as to how well it meets the goal or satisfies the social requirements. A method is developed in Chapter 3 that compares the direct and indirect effects of alternatives. The method permits a direct comparison of the expected internal and external cost differences among a set of alternatives. Where a specific existing system is being modified or replaced, its cost may be used as a reference for evaluating other alternatives.

The final step is the evaluation of the feasibility of implementing the alternative means of achieving a policy goal. This includes temporal feasibility or how long it will require the alternative to reach the policy goal. Thus the temporal feasibility is an expression of the rate of social return of an alternative.

The remaining aspects of feasibility involves technical feasibility, economic feasibility, political feasibility, social acceptability, and timing feasibility. All but the first involves political judgment and are subjective by nature.

The result of such a subjective analysis procedure can become a cost analysis matrix of alternatives measured against a socially derived set of performance criteria if such a result is desired. It is an objective of this book to propose the desirability and method for such a result.

It is proposed that if such analysis establishes the direct and indirect benefits and costs of the set of policy alternatives in dollar (and other resource) terms and provides a time rate of return for each alternative and further provides a measure of the subjective feasibility of each alternative, the decision-maker is then in a position to compare the alternatives within a common framework. It should be clear that the purpose of the evaluation procedure is *not* to determine the single best solution, but rather to provide the policy-maker with a complete assessment to serve as an aid in his decision-making.

The process is not a closed but an open one. It clearly requires a transaction between the political policy-maker and the policy analyst. Any policy analysis procedure is useless if it is not responsive to and guided by the policy-maker. It is he, in the end, who is responsible to the society, and therefore the only one who can decide whether any public policy decision can be made and implemented. Unless the analysis procedure is adapted to his needs it will not be used. Conversely, if the policy-maker is not involved in the evaluation process, decisions will be made not on the basis of social requirements but on the basis of precedent or expediency. The output of the policy analysis process is ideally the selection of the most socially effective means of meeting the social goals or satisfying the social requirements. This decision-making process calls for the selection of a course of action and an allocation of resources for initiating the program.

At this point, executive action begins. Effective executive action is dependent on having a policy goal, the specific approach to reaching that goal, and the resources to implement that approach. The executive process needs to be primarily a technical one to assure that resources are directed to the goals and priorities of the society. It is inevitable that the executive and political functions overlap. The intent of the policy goals is to limit the extent to which the executive branch is required to assign priorities to social goals.

The specific structure of executive organization and programs may be important from an efficiency sense, but are secondary to the process of policy development. The variety of institutional arrangements and mechanical means to carry a public program to completion, any of which might be best in any particular circumstances, are not a subject within the scope of this dicussion. From the policy-maker point-of-view, executive actions and results should be evaluated by performance relative to the same social goals that generated the program, rather than attempt to evaluate organizational efficiency.

Working from this principle, policy-makers can concentrate their efforts on evaluating the performance effectiveness of the program. For example, if it is decided that an important social goal is to increase the safety of highway transportation, it does not or really should not matter to the social policy-maker how the executive branch achieves that goal as long as the administrators do so within the set of constraints established by the policy-maker at the outset. Presumably such constraints would be those that define the limits of the effects of the program on other aspects of highway transportation performance or its indirect effects on other societal systems. The social policy-maker would, of course, specify the safety improvements that were desirable within a defined time period; for example, reduce total number of accidents by one-third in 6 years. In essence, the output of the policy process would be an operational goal and a time table for achieving it. Given this objective function and boundary conditions, the executive agency would then develop an operating program and the detailed resource requirements to achieve the desired goal. Following these steps the executive agency can then develop its program to achieve the objective, its time table for achieving it, and the time table for allocating the resources to do so.

With this degree of definition, the policy-maker is in a position to evaluate the executive program on a continuing basis. For example, at the end of 3 years, if the program has not produced the desired reduction, the program may be discontinued or radically changed. If it is producing the desired results, then at the end of the specified time period, the program can be evaluated again to determine whether it should be considered complete and hence ended or reduced in size with resources sufficient to maintain the target level. In sum, if defined goals are explicit, then the performance of any executive program can be measured and rational judgments made of the effectiveness of executive action. This type of policy evaluation process has been used in industry, especially high technology ones, for many years. It simply had not become a usual or common one in the public policy process and public administration in the early 1970s.

The evaluation phase of policy development is done on a continuous basis for social goals and the effectiveness of policy decisions and resultant executive action relative to those goals. This feedback process permits not only stop or go decisions, but also policy modifications that can increase the likelihood of achieving the desired goals more efficiently and expeditiously. The policy-maker's basic contribution to the process is the information he gets from his own direct contact with the society as applied to his continuous policy analysis process and the evaluation of results produced by the executive agency.

A representative structure for the entire policy process is shown in Figure 2.2. As may be seen, it is composed of three interdependent functions. One is policy-making which is responsible for defining the social goals and requirements representative of the society. Two is policy development or the analysis which transforms the desired goals into alternative operational terms and provides an evaluation of the means for achieving those goals. Three is executive action programming which is responsible for the development and implementation of the means to achieve the policy goal. And four is a feedback that provides the means of evaluating policy as well as program performance.

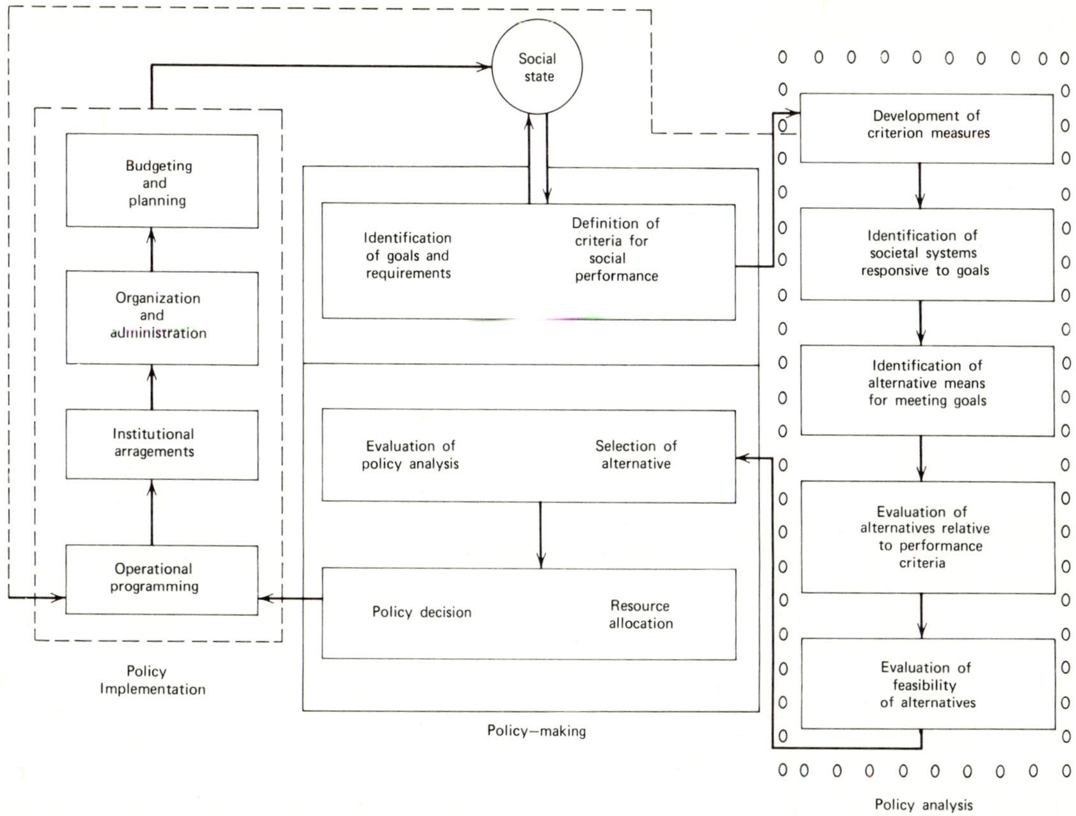

Figure 2.2 The policy process.

GOAL SETTING FOR SOCIETAL SYSTEMS

Societal systems are groups of individuals whose lives and futures are inevitably and inexorably intertwined. Such systems exist whenever two or more persons must be associated and are interdependent for survival or for well-being.

As used herein, societal systems will include all characteristics of the physical and social environment of man. Such systems can range from very small to very large, and the smaller systems are a part of or subsystem of the larger group; that is, they are dependent on and contribute to larger assemblies.

The manner in which societal systems function can be described in many ways. From the beginning of man's existence, these interdependencies began to develop. The result of these interrelations was (1) the sacrifice of complete individual independence in favor of benefits from association, and (2) the need for a way to live together in harmony. The basis for resolving problems of interdependency was some form or another of a political system; that is, a system for resolving differences and for arbitrating competing interests.

In Figure 2.3, societal systems are pictured as controlled by a political system. The nature of this control varies widely, but the primary interest of this discussion is not individual actions, but the collective ones. Accordingly, Figure 2.3 refers to those actions which are within the purview of the political system and the organizational alternatives that are available to a political system. These are alternatives in that changes can be made, provided that the individuals who support the political system concur with the change.

Of particular interest is the fact that *unfulfilled needs* are perceived by individuals. These needs vary widely. Culture, individual differences, and changes in expectations can produce a dynamic condition in most societal systems. The needs are both immediate and long term, but most of the pressure comes from those who press for fulfillment of what is perceived as being needed "now."

For any societal system, individual perception of unfilled needs is an ever-present condition. But also present are the other existing conditions shown in Figure 2.3; namely, resources, population, and culture. The major problems of the political system are (1) to identify the needs (which is not too difficult insofar as immediate needs are concerned), (2) to place priority on those needs, (3) and to minimize or eliminate the needs. Accordingly, political systems take actions in response to the perception of needs.

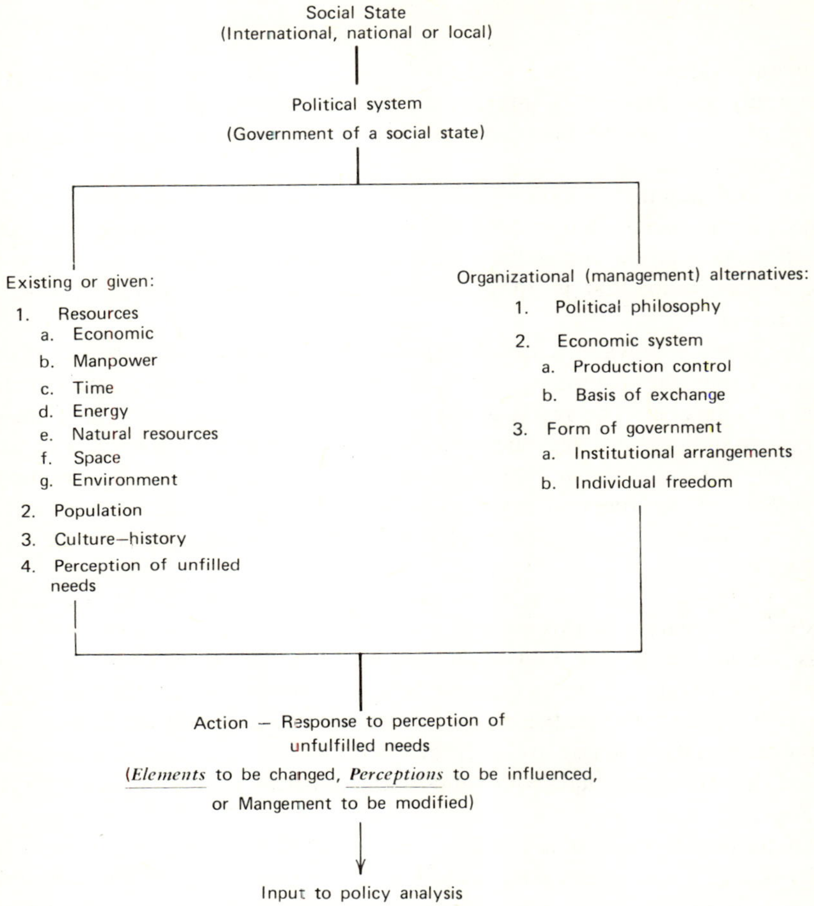

Figure 2.3 Developing input to policy analyses.

Identifying Needs and Establishing Priority

Techniques vary by which needs are identified and decisions made on actions to be taken. The colonial-age town-hall meetings were effective in their day, but became inadequate as the society grew more complex and mobile. Pressure from self-interest groups became a more prevalent method in face of the increased size and complexity.

As a way of obtaining a more rational and comprehensive perspective of individual well-being, the development of *social* indicators received extensive discussion in the 1960s. The complexities and costs of obtaining (and maintaining) a data base were major stumbling blocks, but such a system will undoubtedly be developed and refined.[3] Social indicators are viewed as parallels to economic indicators; that is, they are measures of the "health" of a social system, as economic indicators are measures of the health of an economic system. Table 2.1 contains the list of social indicators included in a 1967 report issued by the federal government.

Pending the development and adoption of social indicators, or a comparable measurement technique, political systems continue to be motivated by pressures from vested interest groups as a means of identifying social needs. This type of decision-making (responding to pressures) aggravates polarization, because for every one who wants to see an improvement (change), there are those who do not want to see change and vice versa. Table 2.1 also contains an example of problems and social indicators that were included in a policy analysis of the airport access problem. With or without an objective method the political system will identify and classify how the ultimate values or ends of the societal system are to be dealt with. Some values cannot be measured on an objective scale (e.g., esthetics). Others do not have a generally accepted standard (e.g., poverty). Nonetheless, some quantitative statement will be needed if these ends are to be useful bases for societal goals. In Chapter 3, possible bases for measurement are described.

Determination of existing conditions of a social need also must be made in a statistically reliable manner. Shortage of data is always a problem, but the comparative variability (time-to-time) and variation (individual-to-individual) must be known with sufficient accuracy to know how the need is to be met.

Responding to Needs

Knowing the needs, and the existing conditions, the problem becomes one of the level of improvement that is to be sought. Two broadly different paths are available. One of these is based on a leader concept in which the societal system is improved in accordance with the views of the leader. The second is a response concept that reacts to the wishes of the majority or the most influential lobby. For either approach, the reaction of the public is important but for different reasons. The leader needs to know how fast change can be made, while the response approach needs to know what is wanted most. Smaller political entities, on occasion, elect to abdicate the

[3] Early in 1974, a new report on social indicators was released by the Office of Management and Budget, Executive Office of the President. This report is discussed in Chapter 9.

Table 2.1 Air Transportation and Social Accounting

1. *Social indicators*[a]	Typical Programs Which Air Transport Can Support
a. Health and illness	Emergency medical services Hospital transport services Safety Air pollution
b. Social mobility	Air transport for low-income groups
c. Physical environment	Air pollution Noise Airport location
d. Income and poverty	Cost reduction in air cargo Job opportunities
e. Public order and safety	Emergency systems Patrol activities Civil defense National defense
f. Learning, science, and art	Student transport Scientific talent transport Research funding Transport of art exhibits
g. Participation and alienation	Air transport for low-income groups Postal services Job opportunities

responsibility of choosing which needs will be served and request professional decisions. This occurs frequently under weak leadership, in small governments with part-time leadership, and/or in the presence of vague and poorly defined perceptions of need.

The first response to the needs is a series of tentative policy goals (Figure 2.4), stated in operational terms, containing the following elements:

1. Reduction of a specific societal need, stated:
   a. Quantitatively (how much).
   b. Qualitatively (in what way).
2. Constraints, including:
   a. Time allowed to produce reduction.
   b. Resources that can be employed.
   c. Limits of impact on social and physical environments.

These first statements can then be analyzed for feasibility. The alterna-

Table 2.1—Continued

2. *Urgent social problems*	Typical Programs Which Air Transport Can Support
a. Poverty	Job opportunities Airport construction and operation Manufacture of airway systems components Aircraft manufacturing Subway construction and operation for airport access
b. Education	Transport of special talents Long-distance transport
c. Racism	Job opportunities Air transport for all income levels
d. Crime	Air patrol systems Emergency systems
e. Safety	Airway systems Aircraft safety Access safety
3. *Special functions*	
a. Emergency services	Helicopter system for highway safety, police, medical requirements
b. Urban transport	Helicopter, V/STOL, and STOL systems for industry and large government installations
c. Postal services	Urban distribution
d. Police and fire services	Patrol and emergency response

Source: Selected Reading 9.2.
[a] Based upon classification system in Selected Reading 2.6.

tives can be examined in terms of (1) facilities, (2) equipment, (3) materials or products, (4) services, and (5) organizational changes. These five categories can be combined in various ways to achieve an ultimate improvement in the societal system. The techniques for analysis are described in Chapter 3. The feasibility of the stated policy goal is thus tested, and if deemed impractical or impossible it is adjusted, reanalyzed, and rechecked for feasibility. This iterative process continues until the political system is satisfied with the operational statement of the goals.

In the practical world of the societal systems of the 1970s the humanistic ends of local and national governments are frequently vague. For a policy

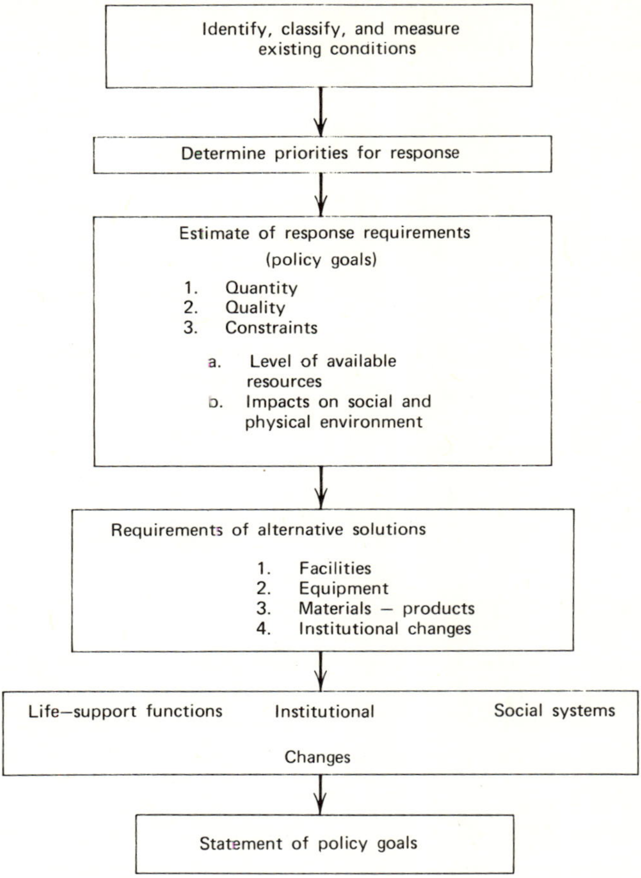

Figure 2.4 Developing response to needs—goal setting activity.

analyst who encounters such conditions, the first step is to elicit from the political system its perception of unfilled needs of the particular societal system. Following this, the priorities for improvement can be elicited.

POLICY GOALS IN A MAN-MADE ENVIRONMENT

The influences of policy on the life of a community were discernible in the United States in the early 1970s, as were the influences of the absence of explicit policies. An understanding of these effects can be obtained by studying how public resources are (1) accumulated and (2) allocated. Thus the amount of resources that become public property, from whom such resources are obtained, and how they are to be used are the crucial issues resolved through policy development.

Various types of taxes, the designation of public lands, and the drafting of personnel into military service have major effects on the life of individuals. The accumulation of such resources by government has both positive and negative effects on individuals, and is, accordingly, a difficult political decision. The benefits, of course, are welcomed, but the adverse effects are resisted. Heavy taxation, regardless of the form, and regardless of the public need for the resources, is not popular.

But without public resources, each individual would need to fend for himself. The extent to which governments provide facilities, equipment, materials, and services varies with the level of government, the wealth of the community, and the current political philosophy. There is no right nor wrong balance in the absolute sense.

The impacts of policy on man's social-economic environment are equally evident. To increase the well-being of individuals, resources are required. Thus economic productivity is a first requirement for betterment of such characteristics as health, recreation, and public safety. Lack of policies controlling economic productivity is a basic principle of free-enterprise, market place economy. Since income distribution when left alone becomes eccentric and because there is human temptation for market manipulation, various degrees of control (policies) generally become desirable to achieve the objective of increased economic productivity and domestic tranquillity. Thus the effects of policy on the lives of individuals are apparent. Both individual neglect and prejudice lead to suffering and hardship of individuals. To intercede, resources of one form or another are then allocated by government.

The role of public policy in man-made environments is thus variable and visible. How policy comes into being, and by whom, may be more obscure but the fact that policy is a dominant factor in man's social, economic, and physical environment is not obscure.

In the absence of new policies, social development occurs under existing policy; for example, the absence of controls on land use or transportation results in unrestrained suburban sprawl. Or with the absence of timely assistance, low income areas and ghettos become pockets of crime.

When public policy evolves as a reaction against undesirable social, economic, or environmental development, the time required to achieve the desired change will always be long. The order of evolving appropriate action is as follows; first, the problem must be perceived; second, a solution must be developed; third, political action must be taken; and fourth, an action program must be implemented.

Most frequently, political strategy becomes enmeshed with policy development, and objective appraisals are difficult to obtain. Polarization then results, with the worst and the best of each solution being exposed, but very little known about the middle ground. Conflict between protagonists with resentment and continued resistance on the part of the losers becomes the undesirable alternative to rational policy development. In such a situation, frequently, the means and the ends become confused and social development is lost sight of as the ultimate goal.

Much of the time loss for response to social need is due to the advisory and competitive practices that have characterized disorganized policy development in the past. Such types of policy development can also lead to serious mis-use of resources: decisions and resource allocations produce too little and too late, or simply too late. Local social change will continue to occur whether there is or is not, new public policy. However, they may not occur in ways that are *constructive* in the aggregate. Because time keeps going, and societies seek ways to grow, traditional policy development practices are invariably late. They tend to deal with yesterday's problems rather than with tomorrow's.

Policy development based on reaction rather than on planned social development is very likely to be working on the wrong problem. On the other hand, planning for social changes during a social transition also poses serious problems. Yesterday's solutions are outdated at a much more rapid pace. Tomorrow's solutions are likely to be treated as heretical and overly liberal.

POLICY GOALS AND TECHNOLOGY UTILIZATION

Defense, space, transportation, and communication technologies were responsive to specific needs in the early part of the twentieth century. Typically the products of these technologies were not rapidly translated into meeting other social needs. As a result, technology utilization did not reach its full potential in assistance in social development.

Resistance to New Technology

Resistance to change is a major factor in the use of new technology. This resistance is felt from the user of the system as well as the designers and operators of that system. The problem is aggravated if the system is relatively large (e.g., a highway or sewage system). Drastic changes for the user may well be impractical.

The problem is further compounded if continued use is made of outdated criteria. Thus if benefit/cost ratios to the user are the criteria for analyzing highway construction, when in reality there is some other policy goal for highways, new technology that could be introduced is resisted.

Standards, codes, and unimaginative engineering are also deterrents to fast utilization of new technology. Standards and codes, of course, are to protect the public from overzealous or unscrupulous industries and from incompetent engineers. Inability to specify and measure performance and to develop an institutional structure to protect against future losses in performance have produced the need for standards and codes, and have permitted incompetent technologists and engineers to continue in practice.

Natural incentives for use of new technology have not existed. In fact, there have been reasons to resist the new. For example, the cost of the research and development or of better engineering usually militates against new improvements and encourages the continued use of existing technology. Certainly the engineering of public works failed to incorporate innovation as a part of the design process. The thrust for adoption of new technology must come from the technologists themselves. The policy-makers cannot be expected to make judgements calling for new technology without back-up analyses from technologists.

What needs to be developed are the ultimate ends (policy goals) by which social development is to be measured. These value judgements must permit broad decisions as to whether a resource allocation *should* be made. Subsequently, they also need to arbitrate *how* the resources should be expended. The value judgement problem is complex at all levels of government, and must be decided independently at each level.

Social priorities, as noted earlier, are a function of the perceived needs of a community, and the political leadership of each community is responsible for this appraisal. The tendency to consider means rather than ends grows out of both the noncommensurate values related to social development, and the special interests that are promoting one function or another. During the midtwentieth century, attempts to use benefit/cost ratio led to some improvement over the preceding studies. But the efforts to convert some social values to dollar terms only served to obscure the view of the policy-makers. Equally disturbing, the desirability of determining whether the activity was economically self-sustaining to the user (or needed to be subsidized) was lost with the assignment of dollar values to nondollar benefits.

The application of welfare economics to policy development has merit of course. This concept is basically one of determining how much one is paying for a specific item of social development. This information, while useful, is not the sole determinant in achieving social development, since some benefits are of higher priority than are others. The perspective that is achieved by assigning priorities to the elements of social development facilitates obtaining, and improves the effectiveness of, the compromises that are inevitable.

The needs for social development are both short-term and long-term in character. Most political systems have not made long-term allocations with gusto because of the short-term needs which are not met. Exceptions to this are construction projects which carry some short-term benefits from employment and sales of materials.

The tendency to "spend for the present" is strongest at the local level. How much should be invested for future years is no less a problem for the political fathers than it is for the fathers of individual families.

Even sleepy little villages will have long-range needs in the form of schools, sewage plants, water resources, and energy needs. The needs will not go away, however, nor will the social development occur in the same way that they would have occurred had the investments been made. Consideration of long-range needs should be preceded by the setting of long-range goals in terms of the accepted set of values for social development. Otherwise, there is no sound basis to consider current investments beyond current needs.

At each higher level of government, consideration of long-range needs becomes more feasible. Far broader tax bases, and the longer-range capability and responsibility permit more consideration of future needs. Once again, however, if means and ends are not separated, the ability to evaluate social development is reduced.

Thus it develops that the policy process is an integral part of the political process and will consist of elements that are required by the political system involved. Policy development, on the other hand, is an analysis from which better decisions become possible. While the policy development process cannot guarantee better political decisions, it should make purely intuitive decisions, or decisions based on primarily political muscle (influence), increasingly more difficult.

Any improvement in political decisions, however, require not only objective analyses, but also a political process that is responsive to the results of such analyses. Two different methods are used in public policy analyses: the social efficiency method and the functional efficiency method.

Social Efficiency Analysis

Social efficiency analyses refer to policy development for achieving *total* social development by analysis of individual well-being, and creating the basic policy goals of a community. That is to say, the analysis starts with consideration of the needs, desires, and aspirations of individuals. The alternative approach, based on *functional efficiency,* also ultimately reduces to consideration of individual well-being, but starts with analyses of functions. Furthermore, functional efficiency creates goals for its own activities and distributes policies from the overall system.

The social efficiency approach is most readily applied to small communities because complexities are multiplied as geographic boundaries and populations are enlarged. On the other hand, the options that are available to small communities are quite limited, and the benefits of the policy analyses are accordingly of lower significance; that is, the margin for error for policy-markers is much smaller.

At state and national levels, a combination of the social efficiency and functional efficiency approaches undoubtedly is most feasible. For professional analysts, the functional approach is most commonly encountered, and provides the most fruitful results. The first step involving how need is perceived, the variability between individuals, and the variation with time is complex, and usually requires the special skills of the behavioral scientist. Economic issues are equally complex, calling for specialists to determine both the availability of resources and the economic impact of alternatives.

Time considerations are social and economic in character, but also heavily political in the way that the rate of change influences the stability of political systems.

The primary output of the preliminary social efficiency study is the range of quantity and quality of improvements that can be sought in individual well-being. The next output is alternative means by which the improvements can be achieved. At this point, guidance is required from the appropriate political system as to the target improvements for which greater detail should be developed. An iterative process ultimately produces the specific policy goals and constraints that are adopted.

The practical difficulties that were present in the early 1970s provided formidable obstacles for the use of the social efficiency approach. These difficulties were related to the state-of-the-art in the behavorial sciences, and to the attitudes and existing practices in the political processes at both the national and local levels. Accordingly, the functional approach provided a better basis for more rational policy development.

Functional Efficiency Analyses

Functional efficiency is based on estimating the impact of changing the quantity and quality of life-support functions. In Figure 2.5, there is a characterization of how the life of an individual can be influenced. As indicated in Figure 2.5, improving the well-being of individuals requires a change in the way that elements are provided by life-support functions. Rather than starting from the improvement desired in individual well-being (as for social efficiency analyses), functional efficiency starts with changing the function and then estimating the effect on individuals.

The functional approach has the disadvantage of being or appearing to be a justification for the expansion (or change) of an existing activity. On the other hand, a body of practical knowledge is available on these functions, and if the analyses are objective and if the results of the analyses of all relevant functions can be converted to the political value system, then a comprehensive view of alternatives becomes available. Failure to include all relevant functions would not produce a comprehensive analysis. Accordingly, the practicality of the approach is limited by the extent to which those several functions can be adequately analyzed.

The functional efficiency approach includes the following three major steps:

- Establish the existing input and output relations for the function.
- Establish functional consequences of alternative changes in output.
- Estimate societal impacts due to changes in functional output.

INPUT-OUTPUT ANALYSIS

For a given function and for a specific government unit, a type of input-output analysis is needed; that is, what results are achieved with the resources used? While inputs and outputs are not static, an annual or other rate can be used to establish the status quo. The conduct of these analyses requires highly competent professionals who are thoroughly familiar with the functional system being analyzed. General categories can be identified, but specific analyses are beyond the scope of this book.

The *inputs* should include all resources that are required by the function. Such resources include (Figure 2.3), but are not limited to, economic, land, materials, natural resources, equipment, manpower, energy, and communication.

The input should be expressed in resource units or dollar terms to assemble comparable summaries on all resources that are consumed. Past investments in the function should be accumulated insofar as practicable and the annual rate of input to sustain the system should be estimated. These input data are needed to develop a comprehensive summary of functional requirements to maintain an existing function at an existing level of output.

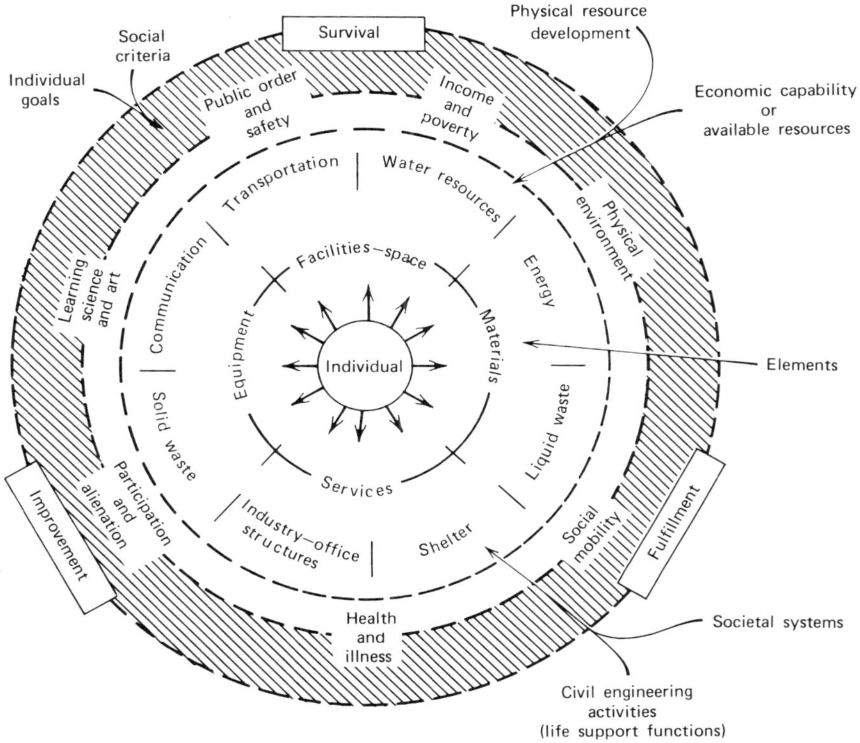

Figure 2.5 Relation between individuals and their pursuits of a better life. Each individual is in search of a "better life." This search involves achieving assurance of (1) survival (food, clothing, shelter, health), (2) improvement (education, training, experience), and (3) fulfillment (culture, recreation, travel). Mankind has progressed through the centuries to the point where his knowledge and thinking capacity precludes his being satisfied with the conditions that he personally can provide or produce. Therefore, the "common" needs and activities require communities of some types. Books for education and recreation; foods and medicine for survival and good health; communication for personal, business and joint welfare; and energy to sustain survival, self-improvement and self-fulfillment produce complex problems. Ultimately, the needs of individuals are for facilities, materials (or products), equipment and services. These, in turn, are used to provide life-support functions. Of primary importance is the development of the physical resources whereby these life-support functions can be provided. But the need for physical resources is not an end in itself—rather, the ultimate goal is the achievement of a better life. The essential and inescapable need for facilities, materials, equipment, and services must not be overlooked. There is no way for an intelligent human race to survive, let alone advance, unless these basic needs are effectively provided.

The output of the function may also be termed the quality of service that is provided. The functional performance is determined first—not the societal impacts. Functional performance consists of quantity and quality characteristics, and can be classed into the following groups for life-support functions:

1. Quantity
2. Quality
   a. Direct Effects
      - Cost to the user
      - Time (or reliability)
      - Safety
      - User Satisfaction
   b. Indirect Effects
      - Physical
      - Psychological

Quantity includes the total output of the function, as well as (1) classes of output and (2) a breakdown of assistance to other functions. To estimate societal impacts, quantity should be disaggregated. Thus cost, time, safety, and such can vary, depending on the function and conditions of use. This type of input-output analysis was not standard procedure in the early 1970s. However, much of the data are available if not always in the form needed. The most difficulty comes in disaggregating characteristics related to the user.

The results of the preceding type of input-output analysis are needed for estimating societal impacts, and do *not* include such factors as the effect on the physical environment. For example, the output of the function would be in terms of quantities of pollutants (physical effects on the environs). The impact is subsequently estimated from a consideration of the pollutant on an areal basis, and on a time scale.

The separation of quality of service and impact has many advantages. First, measuring the output is a technical problem related to the performance of a particular function or system, whereas impacts involve the entire social system. The disaggregations of the output needs to be carried to the detail needed to simplify subsequent estimates of societal impacts.

For smaller communities, or isolated projects, it may be possible to circumvent the analyses of quality of service and estimate societal impacts directly. This short-cut is not as feasible for major system problems wherein components are integrated into a large system (e.g., highways, airports, or flood protection within major watersheds). In these large systems, the problem of whether a *project* should be built is a part of the analysis of whether the *larger system* should be provided.

CONSEQUENCES OF CHANGING THE OUTPUT

As noted earlier, improvements or social development require changes in elements which in turn require changes in the quality of service of a function. In the past this usually required additional equipment or facilities. But beginning in the late 1960s, changes were frequently economically unproductive. For example, the changes necessary to reduce level of pollutants may produce no improvements on the quality of service for the user. Usually, however, functional analyses are directed toward some improvements for the user, either a cost reduction, time savings, or a new degree of safety.

A reduction in cost to the user without an economic subsidy is a reasonable basis for establishing the improvement of a function. The improvement in output of a function may grow out of (1) user complaints or (2) societal concerns. The former has to do primarily with functional efficiency, while the latter is more likely to be related to social efficiency.

Improvements in the quality of service will normally require additional resources, and the input-output analysis can be examined to determine the incremental change in input required for the incremental change in output. This includes both the quantity and quality factors of output, and both the direct and indirect effects of the quality.

ESTIMATING SOCIETAL IMPACTS

To produce a change in the quality of service of a function, changes in input are usually required. In addition, a change in one quantity or quality effect usually produces a change in another. Thus a cost reduction may require a new facility which may become an adverse indirect effect of the new facility.

Given a quantitative estimate of the change in a function, and disaggregated into functional uses, societal impacts can be estimated. The impact estimates that are needed are those previously established as being the measures of social development that are used by the political system. If these have not been established, the problem of evaluating the consequences of a change in a function are magnified. Furthermore, if the values of significance to the political system are not known, the impacts that are estimated may not be those with the highest priority. Accordingly, prior to estimating societal impacts, efforts should be made to elicit the relevant political value system.

The conversion of the change in functional output to societal impact normally requires interdisciplinary studies. The state-of-the-art of this type of estimate was not well developed in the early 1970s, nor were the data readily available in the forms needed. Nonetheless, estimates based on changes in functional output are more reliable than can be otherwise accomplished.

The interactions or multifunctional effects produced by changing only one function can and do produce both positive and negative impacts on social goals. The cost per unit of change, and the total change produced can be used to determine the least costly way to achieve the desired goals. Costs in other than economic terms (pollutants, time, energy, use of space, neighborhood disruption) can be used under a concept of cost per unit of benefit.

From this series of results of cost per unit of benefit (cost effectiveness), a variety of alternatives are available to the decision-maker. In every case, trade-offs are required. If these trade-offs have been effectively developed, the consequences of the decision are well understood.

CRITIQUES, DISSENT, AND CONSTRAINT ANALYSES

Since a policy decision or a stated public goal is a subjective, politically based judgement, it represents one answer out of a group of possible answers. Being political, it is subject to challenge, dissent, and evaluation from the moment that the decision is made.

A political system usually attempts to evaluate the effectiveness of the decision. Political strategy, however, dictates that for some period of time, announcements would emphasize only the positive side of the consequences of the decision. Over longer periods of time, the evaluations lead to adjustments in the policy to account for either lack of effectiveness or changed conditions.

To assure that the consequences of policy decisions are known, critiques, dissent, and evaluations are made by (1) those who wish to know the effect of the implementation, (2) those who are lobbying for a different answer, and (3) those who have a professional or impersonal sense of responsibility for the subject.

The procedures and analyses for dissent follow closely those required for setting policy goals. They are considerably simplified by the fact that the decision is known, and only the consequences need to be determined. This is much simpler than knowing the consequences desired, and studying the alternative means to achieve the desired ends.

As a policy study, a critique can estimate the extent to which the decision may achieve the desired objective. It would also be possible to determine or predict what part of an activity is least effective. To remain a policy study, however, final conclusions cannot be drawn that depend on a political, subjective judgement. The latter is a part of the rhetoric or the political strategy for or against the decision rather than part of a policy analysis.

Objective evaluations are limited, however, if only politically motivated descriptions, rather than policy objectives, have been stated. An example of specific goals were those set by the United States for the maximum level of emissions from automobiles. On the other hand, specific goals for highway safety were not announced during the first 7 years of the program.

The distinction between the technical and the political critique is of greatest importance in a confrontation between two levels of authority. When the superior level has a responsibility for making political judgements, it must do so. Since such judgements are certain to be both supported and rejected, it is critically important for a subordinate level to determine the character and extent of the political and technical judgements. Furthermore, it must be recognized that policy decisions necessarily involve unquantifiable, noncommensurate values. Furthermore, priorities, value-systems, and acceptable bases for describing the welfare of individuals were in a relatively undeveloped state in the early 1970s.

A specialized area of policy analysis is concerned with limits that are placed on individual or group activities. Policy decisions in these areas usually take the form of laws, codes, standards, and regulations.

During the late 1960s, two special forms of constraint analyses emerged: (1) environmental impact and (2) technology assessment. Closely allied with the latter was technology forecasting. These two activities grew out of concern for the degradation of the physical environment. The first produced direct policy decisions to halt pollution. The second was directed to prediction and control of second and third order effects of new technology; that is, attempts were to be made to avoid unexpected side effects. Actions to control pollution or to control or increase the effectiveness of technology became enmeshed in the same set of value system problems that have been discussed. Therefore, the definitions of environmental impact and technology assessment were expanded to include the problem of resource allocation. The term policy development as used herein includes all policy analyses that are used for resource allocation.

SELECTED READINGS

2.1 Baker, R. F., "Engineering Policies, Politics, and Pragmatism," *Transportation Engineering Journal*, American Society of Civil Engineers, Vol. 100, No. TE1, 1974.

2.2 Crocker, T. D., and A. J. Rogers, III, *Environmental Economics*, New York: Dryden Press, 1971.

2.3 Logsden, J. M., *The Decision to Go to the Moon,* Cambridge: MIT Press, 1970.

2.4 Manning, E. C., *A White Paper on Human Resource Development,* Edmonston, Alberta, March 1967.

2.5 Terleckyj, N. E., *Estimates of Possibilities for Improvements in the Quality of Life in the United States, 1971–1981,* National Planning Association, 1973.

2.6 *Toward A Social Report,* U.S. Department of Health, Education, and Welfare, 1969.

2.7 Vanderpool, P., "The Twin Cities Metro Council," *Nation's Cities,* Vol. 10, No. 11, 1972.

CHAPTER 3

ANALYSIS AND SELECTION OF POLICY ALTERNATIVES

In the previous chapter, the identification, specification, and implication of policy goals were discussed. We are concerned in this book with the issue of public policy, and its interaction with science and technology hinges on a specification of goals that has general impact on the society at large.

A general process for policy analysis was provided in Figure 2.1 in Chapter 2, but more detailed generalization of policy development would be helpful. Just as for planning, design, production, maintenance, and operation, the policy development process is fitted to specific applications, and the process varies widely depending on characteristics of the problem.

The procedures that are presented in this chapter are applicable to more narrow decision-making within a system or subsystem, but the main concern is with those larger issues in which considerable uncertainty inheres and for which no detailed or specific technological solutions may yet exist. An example may clarify the distinction. In the face of a serious energy shortage, a society is faced with a problem of deciding on how to allocate resources to eliminate that shortage. Certain alternatives present themselves very clearly as possibilities, such as using energy from hydrogen fusion or converting the temperature gradients of the sea into electricity. Because neither alternative is within the state-of-the-art at present, the issue here is whether investment should be made to make them available. The uncertainties in this situation may be contrasted with those in another example. Should one use colored pavements or traditional pavement markings for the delineation of off-ramps on freeways? Other levels of evaluation questions relative to social, economic, environmental, and political impacts are presented in subsequent chapters.

POLICY EVALUATION

The major concern in this chapter is about the major issues with wide impact. Similarly, it is assumed that at the national level, at least, the public policy-maker is not as much concerned with operations as with goals. The issues that must be decided at this level are those that deal in long-term futures of the society or have significant effects on the society as a whole.

A second consideration is that social goals are not necessarily, or even usually, stated in operational terms. It is one thing to enunciate a policy that there shall be a pollution-free environment. It is another thing to define operationally what constitutes a pollution-free environment. This means that someone must state the number of such things as the parts per million of particulate matter, sulfur or its oxides, carbon monoxide, nitrogen oxides, and such, which will be an acceptable limit for a pollution-free atmosphere. Then, and only then, is it useful to define the means to attain limits, and is it possible to enforce standards on sources, natural or man-made, which may generate polluting agents. In the end, of course, knowing how to state this number is a scientific and technological problem.

Problems arise when the scientist or engineer cannot understand the policy objective well enough to make the operational transformation. Or conversely when the scientist and engineer choose to make policy themselves, which they do usually by telling a policy-maker that a problem exists for which there is only one solution—their own. This is not necessarily a conscious attempt for power, but rather a constriction of a frame of reference due to training or experience.

Policy evaluation is a far broader process than traditional engineering, and it involves a far more sophisticated transaction between policy-makers and science and technology than has historically been the case. On the one hand, the policy-maker must understand that goals need to be formulated in terms that can be dealt with by science and technology; and that there are a multiplicity of means for achieving a policy goal, the best of which have not yet been developed. He must also recognize that any policy alternative has implicit costs as well as benefits. Conversely, the technical community has to provide assistance in making operational transformation of goals, as noted in Chapter 2. Also, it must provide a rational evaluation of all possible alternatives for achieving the policy goals, and must not merely promote parochial technical interests. Finally, it must provide the basic analysis on which a rational policy decision may be made.

Implicit in this discussion is the assumption that a procedure can be de-

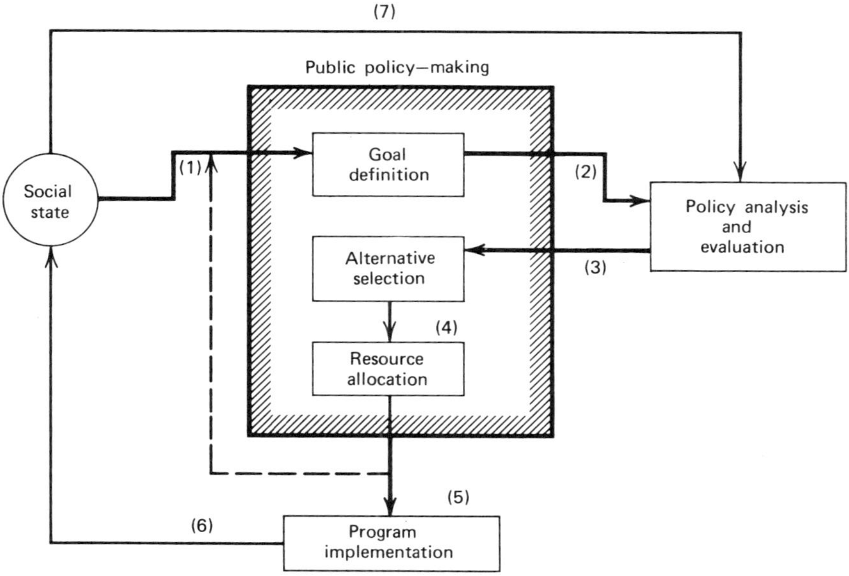

Figure 3.1. Basic steps in the public policy process.

fined that would assist in the evaluation of policy goals. It is apparent that to determine a policy objective three things are necessary (Figure 3.1): (*a*) transformation of the goal into operational terms; (*b*) identification of alternative means for satisfying the objective; and (*c*) evaluation of the effectiveness of the alternative as a means for final selection.

Societal Systems—How They Change

To begin the policy analysis process, the technological or social systems which are sensitive to or determine the means for attaining the social goal or satisfying the social need must be identified. Most social needs and goals arise out of the inadequacies in or the results of the performance of existing social or technical systems. That is, relative to the perceived needs of individuals or groups within the society, the quantity or quality of satisfaction or security is inadequate. This may be manifested in a variety of direct or indirect ways. For example, people may express strong reaction to the perceived level of crime in the society, a reaction that may not be related directly to the objective risk of criminal attack. Rather, it may reflect a general increase in the level of anxiety which is projected onto this class of social process. Conversely, a proposed freeway that is perceived to reduce the cohesiveness of a community through which it will pass may be resisted because of the direct threat that it poses to the members of that community. But in the end, public policy is ultimately responsive, in Western societies, to the *perceived* needs of the culture. Growth and development depend on the responsiveness of the political process to the desired goals of the society and on the ability of the political process to institute changes that express or lead to the attainment of those goals.

The problem for the policy-maker is to determine what changes or additions improve the security and quality of life for the members of the society. In the course of doing this, he is constrained by the systems available for manipulation. There is no way, for example, to obtain higher water quality standards except by treatment or modification of the mechanisms that create and discharge degrading products. In the end, the policy-maker cannot improve the quality of life except by manipulating either the *values of the society* or the *technology used by the society*. Furthermore, the range of choice is limited to a very few manipulable systems. Finally, any changes that may be imposed on any of these systems have concomitant effects both positive and negative on most of the other systems which affect the functioning of a society.

If one examines the systems available for manipulation, 15 classes can be identified. These are listed in Table 3.1. As may be seen, they are organized

Table 3.1 Life Support or Societal Systems

Physical Systems	Natural Process Systems	Social Systems
Shelter	Agriculture	Education
Waste management	Natural resources	Health care
Transportation	Solid	Public safety
Communication	Liquid	Justice
Energy	Gaseous	Protection
Production	Disaster control	Defense
		Welfare
		Recreation
		Culture

into three groups: physical systems, generally under the classification of civil works systems; natural process systems for exploiting or managing the environment for human ends; and social systems for providing order and security.

Within the set of systems there are only a few alternative means for modifying them. Five basic modifiers or agents for changing performance are listed in Table 3.2. One mechanism is *value changes*. In effect, by adopting policies that support certain goals which were not previously acceptable, the social order may be changed. For example, American values

Table 3.2 Individual Needs in the Social State

Elements		
Physical	Services	Modifiers[a]
Facilities	Public	Value changes
Equipment	Private	Institutional rearrangements
Materials—products		Social rewards and punishment
		Actual
		Probable
		Development and adoption of new technology

[a] Modifiers are procedures or agents whereby elements are provided or attitudes are changed.

concerning abortion have changed, so that what was severely punished in the past became a constitutional right. Similarly, because values regarding health care were in the process of changing in the early 1970s, some form of national health insurance appeared likely to be created. This reflects a major change in values about health care and its delivery.

A second modifier is *institutional rearrangements.* Aggregation of productive farm land into large production units is an example. (It may be noted that large units are both profitable and productive in part because of advances in agricultural machinery, which operates most efficiently and economically on economies of scale.) Regionalization of public transportation is another example of institutional rearrangements—attempts to improve the efficiency of transport by aggregating the components to serve a wider and more diffuse population distribution.

A third manipulable dimension is *social rewards and punishment*—laws aimed at changing individual and corporate behavior in ways that are expected to attain a social goal. Antitrust laws are examples of attempts through punishment to protect the larger social interests from *laissez-faire* capitalist interests. Oil depletion allowances are an example of regulations that provide a reward, in this case one which is aimed at stimulating exploration and production of an energy resource.

A fourth variable that may be manipulated is the change of the *probability of reward* for members or groups within the society. Land policies at the end of the nineteenth century are one example. Subsidy programs for stimulating individual activity for personal profit are another example.

The final means for changing the performance of the societal systems is through the development and adoption of *new technology.* As discussed in Chapter 1, technological innovation has been one major vehicle of social change, at least in the Western world. American productivity in agriculture is traceable directly to a massive and long-term investment in research and development and techological innovation in farming methods. American preeminence in electronics and aeronautics is traceable directly to a very high investment in technological innovation. Technology is thus a basic and important means for modifying the social state to attain social goals.

In effect, the policy process involves an intervention in one or more societal systems. The intervention is the institution of some kind of modification in those systems. The modifiers available to the policy-maker are quite limited, probably no more than four classes. Finally, those interventions involve either a physical element or a service element. In sum, the policy maker must (1) decide or choose which system or systems must be changed to attain a social goal; (2) select which modifier will most efficiently attain that goal; and (3) decide what elements within each modifier

Table 3.3 Decision Matrix for the Selection of Candidate Systems[a]

	Individual Influences										
	Modifiers (Agents)					Elements					
			Social Rewards and Punishment			Physical				Services	
Societal Systems	Value Changes	Institutional Arrangements	Actual	Probable	Development and Adoption of New Technology	Facilities	Equipment	Materials	Products	Public	Private
Physical											
Shelter											
Waste management											
Transportation											
Communication											
Energy											
Production											
Natural process											
Agriculture											
Natural resources											
Water											
Mineral											
Disaster control											
Social											
Education											
Health care											
Public safety											
Justice											
Protection											
Defense											

[a] This is a benefit analysis; that is, the benefits which can be derived are to be identified.

will be used to implement the change. Within this context, the policy maker is implicitly or explicitly operating within a decision matrix of the form shown in Table 3.3.

Public Policy Actions—How To Compare Them

The decision matrix, shown in Table 3.3, provides a frame of reference for understanding the domain of public policy. Furthermore, with such a matrix it is possible to evaluate a social or public policy goal and determine what societal system or systems are most sensitive to that goal; that is, which ones will need to be changed to attain most effectively the desired end. For example, if it is decided that a decent home is the right of every American, how may such a goal be achieved? If one assumes for the sake of simplicity that the reason why some people do not have decent housing is because they cannot afford it, several policies can be suggested to make decent housing "affordable." Subsidies can be provided to builders. The technology of housing can be made capital intensive rather than labor intensive. Land can be removed from the private sector and placed under public management. The list can be extended, and almost all of these have been tried in one way or another. (It may be noted that in practice none have been very successful, partly because of how decent housing is defined, partly because of the nature of the market place, and partly because of cultural reasons.)

The kind of matrix shown in Table 3.3 represents a first level of analysis for policy evaluation. It permits the identification of the systems that can be used and the most likely areas for change to attain the desired objective of a social goal. Then the range of possible modifiers may be defined concretely for obtaining the desired change in the societal system; for example, laws, in creating policies on new technology.

This approach provides an entry into policy evaluation. Each of the societal systems is complex. It is essential, therefore, that each of the systems be analyzed carefully.

The required systems analysis is concerned with the functional properties, including the input-output characteristics of the system. The basic understanding of the societal systems in terms of how they perform their essential functions is the only way that a relation may be made between policy goals and system changes. Highway safety represents an example of the complexity of the policy evaluation process. It has been stated as a matter of public policy that traffic accidents shall be reduced. But how can a significant reduction in accidents be brought about? Without an understanding of how the highway transportation system operates and the functional relation between its components, it is not possible to know with assurance what action to take to improve its safeness. If one does not know how the components of the transport system functionally relate, one cannot determine the cause of accidents or performance failure. Consequently, one cannot determine how best to prevent them. In this safety case, the main operational means to attain the policy goals has been to concentrate not on prevention of the collision, but rather on minimization of its consequences. Such an approach appears to be, by its nature, only marginally cost-effective. Congress might have accomplished more by simply passing a law making all state highway departments, all automobile manufacturers, and all traffic engineers legally responsible for traffic accidents. All those institutions would then have upped their investment in research and development from 0.3 to 3.0 per cent. As a result, they would have developed a wholly new set of systems that would have reduced accidents by 90 per cent, and in the process would have produced a level of highway system efficiency such that there would be no need for major highway investments for the rest of the century.

The point of this discussion is that unless the performance of a societal system is understood, it is not possible to predict either the costs or the effectiveness of manipulating any element and hence the likelihood of achieving the policy objective. The fact is that the policy development process has been far too simplistic with the consequence of being at least wasteful and at most destructive of the performance of essential systems.

Changes of Variables within System

What has been said about societal systems as a whole applies equally to variables within each system that may be manipulated to change systems. The variables are evaluated in the same way that the systems themselves are evaluated. Again, highway safety is a good example. The policy was enunciated at the end of the 1920s that driving was a privilege, not a right. Consequently, accidents were deemed negligence and a social punishment was instituted through traffic law enforcement. A very large and complicated institution was developed on the assumption that such a device would act as a deterrent and hence reduce accidents. There is little or no evidence to indicate that this social policy alternative has had or even can have any effect on the traffic accident rate. That is, the assumption of a relation between driver performance and accidents and legal sanctions has been marginal at best. In effect, the alternative means of modifying a system were not carefully evaluated before a policy decision was made. After 40 years of traffic enforcement, the country has come to accept the concept of no-fault insurance. The naive notions of negligence, deterrence, and enforcement die hard. The desire to cling to simplistic solutions even in the face of tremendously complex systems is often overwhelming.

In most cases, the evaluation problem is largely concerned with change agents. Generally, policy is concerned with changes in one class of societal system. Hence the policy process is concerned with the identification of alternative means of modifying a system and evaluating those means. Thus a first step in the policy evaluation process is to identify all feasible change alternatives including those which are theoretically possible, but have not been developed to the point of implementation.

A specific example demonstrates the importance of this identification process. If it is a policy objective to minimize water pollution and it is assumed that this pollution is caused by the discharge of effluents from public and private sources into a water course, what alternatives can be used to eliminate effluents? Before identification of possible policy alternatives, two assumptions simplify the problem. The first assumption is that there are two types of degrading discharge into a water course. One is dangerous materials that are not transformed by natural processes into harmless by-products. The other type of discharge is overload, in that every body of water used for vehicle waste disposal has a maximum quantity of materials that it can biologically transform. The second assumption is that pollutants and pollution levels can be operationally defined and measured. That is, there must be means to measure the pollution level at any time, and there must be means to measure changes in that level so that any alternative implemented can be evaluated, that is, its effectiveness determined.

Table 3.4 offers a series of alternative policies for modifiers for each of the change agents. This table is not comprehensive; however, it does demonstrate the process. One must understand some of the assumptions underlying these alternatives to evaluate their potential effectiveness.

Under institutional rearrangements two alternative policies can be suggested. One is population redistribution. Here the assumption is that the concentration of people on the same effluent sink will cause the input load to exceed the water course's ability to dispose of it naturally. By reducing the population density, it would be possible to reduce the load to the point of being within the handling capacity of the waterway.

The second institutional rearrangement is through land use controls that would limit further loading on the natural environment. Land on or near the water course could be restricted to park or other recreational use.

Under rewards and punishments, at least two possibilities exist. One is to charge polluting agencies an effluent tax. This assumes that either polluters eliminate their pollution to avoid the economic burden, or that the

Table 3.4 Means of Improved Control over a Waste Water System

A. Institutional rearrangement
   1. Population redistribution
   2. Land use control
B. Value changes
   1. Educating people to want to invest in pollution control
   2. Educating industry to want to invest in pollution control
C. Rewards and punishment laws
   1. Effluent charges
   2. Tax incentives for pollution control investments
D. Individual gain probabilities
E. New technology
   1. Reduced costs techniques
   2. Treatment distributed to producing unit
   3. Production process improvements

government use the revenue to provide pollution control technology. Implicitly there is the assumption that the effluent charge will be high enough to motivate action by the polluter and that it is possible to relate quantitatively the tax to the magnitude of the pollutants discharged by the affected agency. It further assumes a largely economic model of human decision-making.

A second alternative in this class is to provide rewards to those who reduce their polluting. An investment credit in pollution control technology could be used. A subsidy for acquisition of pollution control technology is another technique that could be employed. Again it is assumed that the economic incentive would motivate action by the polluter. It is assumed as well that pollution control technology exists which these policies would stimulate polluting agencies to use.

Under value changes would be those policies that would stimulate individuals and groups to act more sensitively toward the environment. It would also include those policies that would motivate people to allocate or reallocate resources to immediate pollution control action. Most of the public relations activities of environmentalists over the past decade have been investments toward this end.

The fourth area, new technology, shows three possibilities. One of the reasons that untreated effluents are discharged is that, for many communities, the cost of collection and treatment is so high that they cannot afford it. This is, of course, the consequence of the technology of waste water collection and treatment. If some new technology were available which significantly reduced the cost, more such facilities could be built. One obvious area for such technological innovation is in tunneling technology, which would eliminate costly hand labor in collector pipe system installation, which accounted for about two-thirds of all the cost of waste water collection and treatment in the early 1970s. The assumption here is that there is a direct relation between facility costs and their installation.

A second kind of technological change would be to shift from massive collection and treatment systems to unit treatment systems; that is, have each pollution-producing unit treat its own. There were at least three technologies in various stages of development that could be so used. One was reverse osmosis; the second was wet oxidation; and the third was waste irradiation. None of these might have been feasible in the 1970s for a whole city, but might have been for an individual home, apartment building, or industrial plant.

A third kind of innovation would be one that increased the efficiencies of the production process. Many pollutants emerge from the process simply because its chemical, thermal, or mechanical efficiencies are very low. If new kinds of processes were developed, the concentration of pollutant discharges could be eliminated or at least markedly reduced.

There is certainly a diversity of possible alternatives for solving such issues as the water pollution problem. This set of alternatives is by no means complete, others could be added to the list. It is necessary, however, to make explicit the assumptions underlying any new alternative's effectiveness.

There is nothing in such a list that allows the policy-maker to make a decision. It is only a first but essential step. Without at least having a comprehensive identification of policy alternative modifiers, the policy process will have less than optimum effectiveness.

CRITERIA FOR EVALUATION

Certainly the identification of alternatives is the first and crucial step in a comprehensive policy analysis process. The *evaluation* of such alternatives, however, cannot proceed until and unless criteria are defined for evaluating them. Effectively, the question is *What measures shall be used to determine the effectiveness of the alternatives in meeting the policy goal?* These measures are policy sensitive; that is, the criteria by which goals are to be evaluated must be acceptable and useful to the policy-maker. But unless the criteria are set, many effects of system changes go undetected and take years to appear. This is especially true with *externalities,* which are the effects of system changes on the other systems or parts of the social structure. For example, it was a government policy for well over 50 years to encourage the development of highways. The long-term implications of this policy never were fully evaluated. But the indirect effects of a policy that consisted of the pursuit of providing universal mobility through highway transportation was enormous. For a total investment of the order of $200 billion, the physical, social, and economic structure of the society was radically altered. It is well, at the outset of this discussion, to review transportation as an example of a technological innovation for meeting a policy goal and to examine some of its more important direct and indirect effects—both beneficial and adverse. This indicates the fundamental importance of a prior analysis of alternatives for meeting a policy goal.

Transportation Policy—A Descriptive Example

There are basically two classes of indirect effects of transportation on the total environment. One involves changes imposed on the physical environment because of the nature of transportation technology. The second consists of derived effects arising out of the operation of that technology.

The first group is fairly obvious. Historically, whenever a transportation system is installed or operated, the physical environment has to be modified to make it operate effectively. The Interstate Highway System is a good example. In this vast construction project, more earth was moved and more natural materials of construction were used than in any other single civil engineering project in the history of man. These modifications in the environment had to be made for safe and efficient performance of the automobile. In essence, the design of the vehicle and its mode of operation dictated highway design. The same, of course, may be observed about airport runways and landing approach patterns, or about railroad track and right of way. The fact is that these changes in the environment, however reasonable, are frequently irreversible; hence an option for use of some other part of the environment is lost.

The other class of effects of transportation is more subtle and far more difficult to separate and define. The development of transportation technology changed the ways in which people lived (see Chapter 1). Some of these changes clearly improved the quality of life while others limited that quality. To get a grasp of the complexity of some of these indirect consequences of transportation, one can examine them on the basis of dominant environmental impact. These may be broken down into four categories: effects on the individual, social effects, economic consequences, and physical effects. A few of the major consequences, both positive and negative, are discussed in each category to provide an insight into the complexities.

EFFECTS OF TRANSPORTATION ON THE INDIVIDUAL

The availability of efficient and economic transportation permitted a significant proportion of the population to enjoy a whole class of opportunities that was never open to them before. Mobility itself exposes the individual to aspects of his physical environment, culture, and history which were not possible previously. Hence a resultant degree of communication among people has leveled many of the provincial differences in the society.

Educational services have been improved radically by the growth in transportation services which has permitted consolidation of facilities and staffs. Health services have been subject to the same effects. The concept of regional medical centers for providing highly specialized services presupposes a transportation capability that is adequate to bring the people to those centers quickly, conveniently, and safely. Exactly the same consequences have accrued for police and fire services. In essence, the mobility provided by transportation in the early 1970s gave the general mass public a sense of independence and individuality that not even the privileged class knew in past eras of human history. This transportation capability, however, also produces negative and disruptive effects such as accidents.

There are also more subtle effects whose short- and long-term physiological and psychological consequences are not clearly understood. What are the long-term effects of moderate levels of noise, shock, and vibration? That they are annoyances and that they can generate stress is known. Too little is yet understood about the human tolerances to repeated, random exposure to project the effects accurately. In the development of newer high-speed surface transportation, the problem of shock, vibration, and noise caused by vehicles traveling at speeds of 150 to 300 miles per hour can be expected to become more significant and predictable.

Air pollutants and their effects on human health have been amply documented. Transportation systems which use fossil fuels as energy sources become sources of air contamination.

SOCIAL EFFECTS OF TRANSPORTATION

Transportation also changed the character of the social environment. The rapid urbanization that occurred in both post-World War eras was another result of the large investment in highway transportation. The real growth of urban population came not in the center city but in the surrounding suburbs. What emerged was a complex social structure classified as the metropolitan area. Ill-defined communities developed with looser associations among groups than any society had ever known. What developed was a large increase in informal or voluntary subgroups which existed to meet a range of social and personal goals that could not be achieved without an instantly available, multidirectional, personal transportation capability. Hence transportation provided a means by which a variety of social institutions could flourish and provided opportunities for masses of people to experience the largest range of cultural activities any society had known.

Mobility also decreases the durability of some groups. Institutions, stable and constant for years, change character rapidly. This occurs in part simply because high mobility causes a decrease in continuity of community membership. People are freer to move and were moving on the average of every 5 years (in the early 1970s). Consequently, community and social institutions take on different meaning. Family structure and friendship relations change significantly. As a result, according to some sociologists, there is a decrease in the sense of stability and permanence in the culture. Again, transportation technology plays a significant part in initiating these changes, but the degree and ultimate effects on the social environment are not adequately understood—much less evaluated in policy-making terms.

Negative social effects of transportation also include the outward migration from city centers which reduces the tax base that cities need to meet their problems. The corollary result is *de facto* segregation of the urban poor from the affluent suburbs.

Yet as the major cities lose their commercial industrial base, they wish to retain and even expand their communication, command, and control functions. These functions require the large masses of workers and professionals located in the suburbs to converge daily in a wave to the city center. This leads inevitably to a new group of congestion, parking, and pollution problems. Improved transportation technology thus tends to compound urban problems—even the problem of urban transportation.

ECONOMIC EFFECTS OF TRANSPORTATION

There are significant economic effects from transportation technology. It is clear that the mass economy enjoyed in the United States was made possible in significant part by innovations in transportation technology. First railroads, then highways, and then air transport brought about the expansion in the range of products and services available by reducing sharply the costs of movements of goods and people. The cost of distribution (transportation) has always helped determine where industry and warehousing was located. Improvements made in transportation technology typically increase the flexibility in plant location. The cost of transportation became so marginal that the choice of plant location was increasingly based on market, labor, tax, and social criteria rather than distribution (transportation) costs.

Thus it developed that by 1960 one key consideration underlying the highway construction policy-making of the Appalachian program was that transportation technology provided a basic infrastructure needed for an isolated area to develop economically. This had become a consistent finding of economic impact studies. Land values around highways and new airports have all risen, and appropriate business activity has been generated. Certainly residential housing followed the growth and improvement in highway facilities. The mobility provided by transportation facilities to economically underdeveloped areas made large areas accessible and attractive for living, a fact which was quickly exploited by business and home builders.

The location of transportation facilities not only stimulates economic activity, but also tends to change it somewhat. Highway by-pass studies, for example, indicated that businesses adversely affected after such a change in a transportation facility were those which catered directly to the highway user. Similarly, land values tended to rise in the area of an airport. But the kinds of economic activities that the land can support have been determined frequently by characteristics of aircraft activity.

PHYSICAL EFFECTS OF TRANSPORTATION

In addition to the individual, social, and economic effects that derive from transportation, there are also significant physical consequences that arise. Whenever a highway, airport, or railroad is built, the local ecology is changed. The construction of highways can change the hydrologic properties of streams and watersheds. Changes in the rate and direction of water flow have subtle effects on stream characteristics as well as on erosion.

The location and operation of a transportation facility can add also to the pollution problem. Air pollution has been mentioned, but there is also water and soil pollution deriving from the construction and operation of transportation systems. Heavy construction can lead to the silting of streams, which affects their purity as well as affecting fish population.

To improve all-weather operation for air transport, methods for dispersing of fog or the reduction of snowfall are very attractive. Cloud seeding and other weather modification techniques could provide significant improvement in air terminal and air traffic control operation. Yet the effects of any weather modification may extend far beyond the local air space. Hence a very clear understanding of the consequences must be had before large scale operations are undertaken.

Defining Criteria

Many of the effects on the environment enumerated in the preceding transportation example were wholly unpredictable, but a very large number of them could have been anticipated. Certainly, the pollution space boundaries for all modes of transportation could have been predicted. By the 1930s, enough was known about exhaust emissions to predict what numbers and concentrations of vehicles under what atmospheric conditions would produce significant levels of combustion by-products, but the prediction was not made. Certainly the accuracy of such a prediction would not have been high, given the state of knowledge at that time, but at least it would have given the policy-maker considerable warning of what might happen. Then a monitoring system could have been placed in operation so that the panic measures that were required might not have become necessary. The same is true for several other significant externalities such as noise, congestion, and energy considerations. Since the technical community was not pressed (and did not volunteer) to provide such analyses, most of the major consequences of the implementation of alternatives went unevaluated or were evaluated in a limited, parochial, and shortsighted way.

The criteria for social goals are discussed in other chapters. The use of social indicators or critical problems as measures of quality of life are helpful if they are responsive to the frame of reference of the policy-maker. Table 3.5 is the classification used in this book.

Such evaluation hinges on the availability of a set of criteria against which each alternative may be evaluated. These criteria fall into three general classes: *environmental criteria*; *resource criteria*; and *social criteria*. As we see, the technical problem is not that of identifying the criteria, but

Table 3.5 Classes of Criteria in Public Policy-Making

Subject Criteria Classes			
Environmental	Resources	Social	Problem Criteria Classes
Pollution	Materials	Safety	Poverty
Esthetic quality	Energy	Public health	Education
Land use	Time	Shelter	Racism
Biotic system	Money	Mental health	Crime
	Manpower	Ethnic quality	Pollution

one of defining the metric or standard of measurement for them. A listing of the major criteria grouped by class is shown in Table 3.5. Each of these must be defined explicitly.

Under *environmental criteria,* the major classes are labels for the direct and indirect effects of any proposed change in a societal system. One of these is *physical pollution*—either solid, liquid, or gaseous. Things are pollutants if, and only if, they temporarily or permanently reduce the viability of the physical environment. The variety of pollutants is enormous and most, but not all, are by-products of social operations. The existence of pollutants normally arises out of a public policy decision, as an externality of that decision. For example, high performance aircraft were developed as a result of national defense policy. The sonic boom, an energy pollutant, is an externality of that policy process. Similarly, a local land use decision, for example, industrial parks, can bring about a concentration of industry that as a by-product generates a significant increase in air or water pollutants. In any case, pollution represents a basic dimension for evaluating any policy that implies social change.

A second subject class of environmental analysis is *land use.* Changes in the use of land may change markedly most or all other environmental qualities. Policy alternatives that involve zoning for physical development or development priorities cause modifications in the environment. Federally insured housing is an example of a public policy alternative that drastically changed the use and value of land, especially on the periphery of major cities, and resulted in establishing the suburban environment.

A third environmental subject class is *esthetic quality.* The restrictive public housing policy that forces high rise and unimaginative structural design can create an esthetically as well as a socially dissatisfying environment. Similar consequences can occur with the construction of urban freeways. In general, esthetic quality is a criterion in public policy-making over which considerable controversy has developed, largely because of differing definitions. However, it is generally accepted as a legitimate environmental concern.

A fourth environmental consideration is that involving our *biotic* systems. The public turmoil before, during, and after the challenge of conservationists to the continued use of DDT is an example here. The variety and complexities of the biotic systems which are part of our environment are nearly infinite and most often affect the quality of health maintenance, including nutrition.

Resource criteria are the second of the major groups. These also may be categorized into classes. The first is *material resources,* including the raw materials, recoverable and nonrecoverable, out of which the social systems are fabricated and their artifacts created.

A second resource criterion class is *energy*. These resources include all basic elements that can be converted into useful energy for driving man-made systems of production and consumption. Historically, energy sources have been combinations of recoverable and nonrecoverable resources.

A third type of resource criterion is *time*. To a large degree, time is a human resource. It is related on the one hand to the time required from a unit of labor to create a unit of production. On the other hand, it is the time required by an individual to achieve needed satisfaction. Thus time is both a production resource and an opportunity resource. As will be seen, time becomes a fundamental and significant social efficiency criterion and its loss becomes a qualitative as well as a quantitative cost.

A fourth class of resource is *money*. In this context money is best viewed as a token representing "stored labor." It is a value—or resource—only to the extent that it is exchangeable for the production of the social system. Money, therefore, is to be allocated consistent with policy, as with any other resource.

The fifth class of resource is *manpower*. This is a finite resource which is inherent to the social or governing structure and which has both quantitative and qualitative limits, which are controllable variables dependent somewhat on the status of the nation's socioeconomic and technological development.

Social or human criteria are the third group of criteria. Five classes are identified. The first is *physical safety* or the damage or destruction to members of the society caused by the unpredictable operation of natural and man-made systems. The concern here is with forces ranging from earthquakes and floods to industrial and transportation accidents. The source of occurrence of these events shifted markedly during the twentieth century from the natural systems to the man-made technological systems.

The second is *public health*. This refers to the state of disease or physiological disorder within the society. It would include the type, frequency, and severity of infectious and degenerative disease processes within the society.

A third type of social criteria is *shelter quality*, in which the concern is with the extent of protection offered the individual or family unit from the normal range of environmental forces. It would include not only the physical structure but also the facilities needed for satisfying the basic security needs that are contained within the structure.

The fourth type is *mental health*. This may be defined in terms of the level of personal integration of individuals and the cohesiveness of groups within the society. To a large degree, mental health relates to the amount of energy required or expended in maintaining personality integration as opposed to that expended in production, achievement, or creation.

The last type of social resource criterion is *ethnic quality*. This is concerned with disparities in level and quality of rewards and satisfactions due to racial or ethnic differences. It is assumed that such disparities derive from environmental restrictions and social sanctions imposed on such groups. Ethnic quality thus reflects the differential discrimination among identifiable groups within a society.

The criteria that have been defined are neither all-inclusive nor were they generally accepted in the early 1970s. However, they do provide a comprehensive and useful point of departure. Furthermore, they cover the range of values that are of concern to the American society. A policy that may demonstrably increase or improve the quality along any of these dimensions is considered a social good, all other things being equal.

Measurement Scales for the Criteria Classes

As a first step in the policy evaluation process, comparisons are made of all alternative policy options relative to the preceding criteria classes. The first step is to identify those criterion dimensions that have an impact on the alternatives under consideration. Any set of alternatives may not change or impact all the criteria, and some alternatives impact one combination while others impact a different combination. Within this framework, it is possible to define a first-order matrix of the policy alternatives and the evaluation criteria. Such a matrix allows the policy analyst to identify the social criteria that are influenced by each alternative. Using the earlier example of water pollution control alternatives produces the matrix for modifiers shown in Table 3.6 for each of the given alternatives.

For present purposes, three ratings were used—definite impact, possible impact, and negligible impact. As may be seen in Table 3.6 there are major differences among the alternatives in terms of their impact. However, across the whole set of policy options, every criterion variable is affected either directly or indirectly. Thus all the alternatives will change water quality, but major population redistribution will also have an impact on the social state criteria, whereas educational efforts will not. In sum, as a first step in the policy evaluation process we have identified the possible alternative means of attaining a policy objective. We then have identified the criteria for evaluating those alternatives. And finally, we have related the policy alternatives to the evaluation criteria.

This first step of the analysis is wholly qualitative. It says nothing about the magnitudes of the direct and indirect effects of any policy. It is essential that a decision process find a way to measure the differences in effectiveness of each policy option. This, in turn, requires the development of

Table 3.6 Description of Sensitivity of Social Criteria to Water Quality Policy Alternatives[a]

	Policy Alternative								
	Institutional Rearrangement		Value Changes		Laws		Technology		
Criteria Classes	Population Redistribution	Land Use Control	Education of Individuals	Education of Industry	Effluent Charges	Tax Incentives	Cost Reduction Technology	Distributed Treatment	Production Process Improvement
Environmental									
Pollution	√	√	√	√	√	√	√	√	√
Esthetic quality	√	√	?	?	√	√	√	√	—
Land use	√	√	—	—	?	√	√	√	?
Biotic	—	—	—	—	—	—	—	—	—
Resources									
Materials	√	—	—	—	—	√	√	√	√
Energy	√	—	—	—	√	√	√	√	√
Time	√	—	—	—	?	?	√	√	?
Money	—	—	—	√	—	—	√	√	√
Manpower	√	—	√	—	—	—	—	√	—
Social									
Safety	—	—	—	—	—	—	—	—	√
Public health	√	√	√	√	√	√	√	√	√
Shelter	√	√	—	—	—	—	—	√	—
Mental health	√	—	—	—	—	—	—	—	—
Ethnic quality	√	—	—	—	?	√	—	?	√

[a] √ = definite impact
? = possible impact
— = negligible impact

some metric for each of the criterion dimensions against which each alternative may be measured.

There are two ways of accomplishing this quantification. One is to transform each criterion measure to a common objective scale. The other is to develop a common subjective scale for each which has the appropriate mathematical properties. An example of the former would be a scale of dollars. An example of the latter would be a psychometric scale of utility.

Before developing either, it is important to develop some elementary ground rules for measures. It is assumed that for policy evaluation, it is necessary to know how much better, or more effective, one option is than another. This presumes that there is a measurable "distance" between the alternatives. This then assumes that a metric may be defined that has equal intervals. An example may clarify the problem. An interval scale is one which has the property such that the distance between any successive units are the same. A temperature scale is an example in which the difference between 20 and 21 degrees is the same as between 80 and 81 degrees. Similarly, the distance on the scale between 20 and 60 degrees is the same as between 50 and 90 degrees. Thus interval scale has order and distance as its essential mathematical properties. It does not have a zero point, or if it does it is merely arbitrary. Zero degrees on the Fahrenheit scale has no unique meaning; it is merely one of the ordered set of numbers on the scale.

In general, an interval scale is the minimum measuring device, for it at least allows the operations of addition and subtraction to be performed. For purposes of policy evaluation, where it is necessary to know how much more or less effective one alternative may be than another, this scale has the minimum mathematical properties needed. It also, incidentally, provides the minimum properties required for the application of normal parametric statistical techniques of analysis.

It should be noted that it is frequently possible to develop an interval scale from simple ordinal data, that is, rank order data. A wide range of methods for both developing and testing such scales was developed in psychometrics between 1930 and 1970.

Ideally, one would like to have a measuring instrument that permitted the operations of not only addition and subtraction, but also multiplication and division. This requires the development of a ratio scale, one that has a true zero point. An example of such a scale is a scale of mass in which not only is 4 grams 2 units greater than 2 grams, it is also twice the mass. In social measurement, several techniques have been developed for constructing such ratio scales. These may be used to develop metrics for qualitative or subjective attributes.

The measure problem for policy evaluation is thus a complex one. This is especially true for those criterion dimensions that are basically qualitative in nature. There had been little funding of efforts to scale values explicitly or to use such metrics for social decision-making by the early 1970s. In part this lack was because security oriented societies were engrossed in physical control activities so that these domains were the ones to be quantified. Also values of a philosophical or social kind have been treated as global abstractions. They were not defined in operational terms so that measures could be derived for them.

It should also be pointed out that each of the criterion dimensions may already have a measure attached to it. For example, pollution is measured along several dimensions of the concentrations of polluting elements. Safety is quantified in terms of numbers of accidents relative to time, population, and such. However, esthetic quality and shelter quality do not have objective measures. Furthermore, each of the dimensions has, if it has a measure at all, a unique metric so that one cannot be combined in any mathematically rational way with any of the others.

In general, there are two presently feasible solutions to the problem. One is to use a "mixed metric" and evaluate the alternatives by a simple profile analysis. At the very least, it should be possible to define the change in any of the criteria in its own common units. If an alternative takes land out of productive use, it can be defined in terms of acres or square miles. If it reduces energy requirements, this can be defined in terms of BTU/time (British thermal units). Safety can be defined in terms of numbers of accidents per 100,000 population. Some of the criterion measures are hard to come by, but dimensionality can be found for all.

Another solution is to use a subjective scaling technique. Here, each criterion dimension is used to measure subjectively the value of each modifying alternative; for example, a nine-point scale ranging from significantly degrade to significantly improve social performance is the criterion dimension. The scale would look like the following:

— −4 — −3 — −2 — −1 — 0 — +1 — +2 — +3 — +4—

Significantly degrade	Neither degrade nor improve	Significantly improve

If (1) the ratings can be shown to be true interval scales, (2) the variance in ratings among the scales is not significantly different, and (3) the individual criterion metrics are linearly transformed to a common scale, it is legitimate to sum the scale values used in the impact matrix; then the column sums represent the best *estimate* of the overall effect of each policy alternative on the criterion dimensions. In Table 3.7, the scale value for each alternative for each criterion has been entered as an example. The column

Table 3.7 Sensitivity of Social Criteria to Water Quality Policy Alternatives[a]

Criteria Classes	Policy Alternative								
	Institutional Rearrangement		Value Changes		Laws		Technology		
	Population Redistribution	Land Use Control	Education of Individuals	Education of Industry	Effluent Charges	Tax Incentives	Cost Reduction Technology	Distributed Treatment	Production Process Improvement
Environmental									
Pollution	+2	+1	+1	+1	+3	+2	+3	+4	+2
Esthetic quality	−2	+1	+1	+1	+2	+1	+2	+2	+1
Land use	−4	+1	0	0	+1	+1	0	0	0
Biotic	0	+1	0	0	+1	+1	+2	+2	0

Resources									
Materials	−2	0	+1	+1	+1	+1	+3	+3	+1
Energy	−4	0	0	−1	−1	−1	−1	−2	−2
Time	−3	0	0	0	−1	−1	+1	−1	−1
Money	—	—	—	—	—	—	—	—	—
Manpower	—	—	—	—	—	—	—	—	—
Social									
Safety	−3	0	0	0	0	0	+1	+2	+1
Public health	−1	0	0	0	0	0	0	0	0
Shelter	0	0	0	0	+1	+1	+2	+3	+1
Mental health	−2	0	0	0	0	0	0	0	0
Ethnic quality	−1	0	0	0	0	+1	0	0	+1
TOTAL	−20	4	3	2	7	6	13	13	4

[a] +4 = significantly improve
0 = neither degrade nor improve
−4 = significantly degrade.

totals represent the sum of these estimates. Assume that the three criteria for measure are met. Also let us assume that the variance is unity. Then we can say that a one-unit difference in these column totals is a significant difference. Thus the first two technological alternatives are better policy alternatives than are any of the others. It may be noted in passing that population redistribution is a negative alternative in the sense that it is perceived to produce significant negative effects on the total quality of the physical and social environment.

This overall approach to policy evaluation must be applied with caution. First, these ratings must not be made in ignorance. Someone who understands the nature and performance of the policy alternatives must provide the estimates. This does not necessarily mean "experts" or technocrats. It does mean people who have thoroughly familiarized themselves with each alternative and how it would function.

Second, the statistical assumptions underlying the scaling procedure must be validated in any application. Aside from testing the internal consistency of scales produced, tests for the validity of the additive assumptions should also be made. Given several independent estimates, the means and variances of the row and column sums can be calculated. Using standard analysis of variance methods, the reliability of judgments between raters can be determined. If there is no homogeneity of variance, then simple rank tests may be used to determine differences among alternatives.

Third, the scaling assumptions employed in this method are simplistic. Not only is it assumed that the ratings are on a direct interval scale, it also is assumed that the criterion dimensions, or the rating of them, are independent. This is a marginal assumption for which no proof may be given. It is, however, a reasonable starting point and one that can be considerably strengthened by using appropriate univariate or multivariate scaling techniques.

The method developed here is general and can be applied to any set of policy alternatives defined to achieve a social goal. It permits the policy analyst to evaluate those alternatives against a common set of social criteria and identify those that offer the highest promise of producing significant social gains. As should be obvious, the method can be used to evaluate policy alternatives in the social as well as physical area. If the alternatives are reasonably defined and their function and performance reasonably explicit, the evaluation procedure should produce reliable results. As a matter of fact, if the composite ratings have high variability among qualified raters it is a good indication either that the alternative has not been defined precisely enough, or that there is no way to estimate its impact on the criterion. The policy evaluation process would then give the policy-maker an estimate of the unreliability of the prediction of effectiveness of an alternative, as well as of the desirability of further study before establishing policy.

COST EFFECTIVENESS EVALUATION

Up to this point, the approach has been aimed at identifying and evaluating those possible options which provide maximum benefits in solving a problem in the public domain. The proposed method also permits the explicit identification of externalities of those alternatives. That is, it identifies the possible impact of any alternative on societal systems other than the one to which it is addressed. In the ideal case, where all other things are equal, the policy-maker would always choose that option that satisfied the need or requirement of the specific system and had no negative effect on any other system. The ideal rarely exists because changes in one societal system always generate changes in others. The best one can do is to maximize the effectiveness of a change in one system and minimize its negative effects on as many as possible of the other societal systems. However, there must be a means for measuring the impact, the internalities as well as the externalities, of any of the set of policy alternatives. The question becomes one of deciding how that process of measurement will be accomplished.

Basically, there are two ways of approaching this kind of impact analysis. One way is to measure each identifiable effect, internal and external, in its own criteria units. This approach has been used for quantifying the social and environmental criteria evaluation for alternative highway route locations. For example, it is possible to determine for each alternative route the number of people displaced, the acreage of open space taken, the dollar volume of businesses removed, the noise levels generated, the carbon monoxide levels produced at design capacity, and so on. One then has the highest "score" for that alternative which minimizes the negative effects and produces the largest traffic benefit, measured, say, in travel time savings.

The difficulty with this kind of approach is that there is no scale of quality relative to each of these dimensions. For example, how much more negative is the taking of 2 acres of parkland than 1 acre of parkland? Is the displacing of 1000 families 10 times worse than the displacement of 100 families? But what is more difficult with this approach is the trade-offs. That is, how much more negative an effect is it to take 5 acres of parkland than an extra 100 homes? Or worse yet, how much more negative is it to take 1000 homes and 1 acre of park than 500 homes and 10 acres of park? There are no quantitatively acceptable answers to questions of this type.

Yet they have been made implicitly by policy-makers for many centuries, most of the time without their realizing it.

It is well to point out, all that this demonstrates is that the public works engineer and policy-maker must now consider criteria other than geography, geology, and engineering economics which do have measures that may be combined, that is, total dollar cost. Where engineering has always followed the path of least physical resistance, it must, in addition, now follow the path of least social resistance. It is clearly not so obvious a path.

A second approach to the impact analysis is to find a way to place all effects on a common measure continuum. If such a single measure can be developed, then there will be commensurability across all alternatives. The values of the impact on all criterion dimensions may be added to give a simple overall measure of the effect of implementing any alternative policy. At that point it is possible to compare directly and quantitatively all the alternatives. One of the most obvious ways of doing this is by using a scale of dollars as the metric. This may seem surprising on the surface, but both practically and psychologically, the dollar scale is one of the few yardsticks which the society as a whole can understand. In effect, it appears to be an implicit measure of value. In the end, its utility for policy analysis purposes depends on how it is used.

It would appear that dollars as a metric has the same properties as most subjective scales. That is, it is elastic and relative. It does not appear to have any extrinsic, absolute value. For example, it is not likely that the society could agree on a dollar value for a human life. Is each and everyone worth $1000, $1 million or $1 billion? Such a question implies that we have an absolute scale of value. The fact is that no such scale does or can exist. However, if the dollar scale is used relatively, the decision process generally becomes obvious and direct. Suppose there are two alternative means for saving the same number of human lives; one costs $10 billion and the other costs $1 billion. Assuming no externalities, there would be universal agreement to implement the second alternative. Given a policy goal then and a series of alternative means to achieve that goal, if the total direct and indirect dollar costs for each are determined, immediately one can decide which alternative is most cost effective.

In the example used, the effectiveness was set at the same value for all alternatives. Thus only the total cost differences are needed to make a policy decision. It is obvious that the effectiveness of the alternatives may differ as well as their costs. Then it will be necessary to evaluate both jointly. That is, it becomes necessary to carry out a trade-off analysis. As shown later in this chapter, by combining all positive and negative effects on a dollar scale, a large portion of this class of problem can be avoided.

There is an additional requirement for a useful economic evaluation of policy alternatives. This is that acceptable levels of performance may be defined for any of the societal systems that the set of policy alternatives may impact. We have to know at least what minimum performance levels we are willing to accept. At the very least, present performance levels may serve as such a reference. For example, if an alternative that improves the performance of one system also degrades present performance of another by say 10 per cent, then we can say that the second system must be compensated by that amount to return it to its present performance level. If the unit cost of that system is known, then it is possible to calculate the costs to return that system's performance to its acceptable level. In terms of policy evaluation, this means that the alternative that induced that added cost must have that increment added to it. As is obvious, this kind of a procedure permits a simple algebraic addition of the costs across systems. For example, let us suppose a ban on all vehicular traffic into city centers is a policy alternative. There would be a cost associated with this due to the increased travel time for persons traveling to the city. In addition, such a policy would significantly affect construction within the city for there would be no way to bring in materials of construction. This loss may be estimated and hence the total cost of the alternative computed. One could contrast this alternative to one in which all vehicular travel into the city center would be underground except for construction vehicles. For the moment assume no other externalities. Now the only cost would be directly related to the cost of the subterranean highway facilities and parking. One can estimate the costs of this policy alternative and make a direct comparison with the other. The essence of this methodology is to cost out direct and indirect effects of any policy alternative.

Cost Evaluation Model

The logic for the cost effectiveness of a set of alternative ways to achieve a policy goal may be stated quite simply. For each alternative, there are two sets of dollar costs associated with its implementation. One set may be classed as direct costs. The other set may be classed as indirect costs. Since all of them are scaled in dollar terms, they are at least algebraically additive. Hence a figure of merit for each alternative may be calculated which is the algebraic sum of its equivalent dollar costs and benefits.

In general, the goal is then to estimate the equivalent costs for each indirect element for every alternative that is a member of the evaluation set. Each of these elements is treated independently. That is, for policy purposes it may be assumed that a change in any one element does not affect the cost of any other element. Hence the model is essentially an independent one, and with a couple of exceptions, it is assumed to be a simple linear one. It is a model of the form:

$$
\begin{aligned}
C_{A_1} &= a_{11}X_1 + a_{12}X_2 + a_{13}X_3 + \cdots + a_{1n}X_n \\
C_{A_2} &= a_{21}X_1 + a_{22}X_2 + a_{23}X_3 + \cdots + a_{2n}X_n \\
&\vdots \\
C_{A_k} &= a_{k1}X_1 + a_{k2}X_2 + a_{k3}X_3 + \cdots + a_{kn}X_n
\end{aligned}
$$

where C = the total cost of the alternative
X = the cost of one element
a = a weight for the cost element

For the purpose of policy alternative evaluation it is also assumed that the initial weighting coefficients are unity. Since, at the outset, most of the economic calculations are of the scouting variety, it is not necessary to weight the cost differentially. Doing so is perfectly feasible, but there is no obvious way to define the functional form of the coefficients. Consequently, it poses serious mathematical difficulties and can make the solution of the set of equations impossible. It is preferable to leave any weighting to the end of the evaluation.

To apply the model, it is essential to define the elements for both direct and indirect cost classes. Each are defined in this section. First we define the element and then a metric for it.

Under direct costs there are seven components that should be considered as cost sources. These are the costs for:

1. Research and development
2. Engineering
3. Capital improvements
4. Materials
5. Maintenance
6. Management
7. Labor

The variable costs of production (numbered four through seven) are actually annual costs while the capital cost components (numbered one through three) are one-time costs. In normal economic analysis of the production function, all costs would be annualized in some relation to demand. However, for the purpose of *relative* evaluation of public policy alternatives, they may be treated separately. That is, research and development, engineering, and capital investment may be treated as a variable fixed cost while operating costs would be calculated on a demand based annual function.

Research and development costs as used here include all those costs in manpower, facilities, and materials required to create the system design that will meet social requirements. Not all policy alternatives require research and development.

Engineering costs are those required to transform the system design into an operational system. In defense system parlance this is advanced development. A specific example is the case of the air bag restraint system for automobile accident injury reduction. Considerable investment was made in research and development of the design of such a system. It involved the analysis and development of sensors, triggers, rates of inflation and deflation, materials, and packaging. However, to transform these designs into a mass-produced product is a large, complex, and costly engineering undertaking.

Capital costs are essentially those resources required to produce in place a functional system that provides the societal benefits of the alternative. In public works systems these are almost always tax dollars. Occasionally these may be private investments, such as in toll highways. However, under either circumstance, it is a capital investment required by a public policy decision.

The variable costs associated with any alternative are the standard ones. They include materials, labor, management, maintenance, and administrative costs. They may be calculated in this analysis on an annualized basis. For policy evaluation purposes, this is generally sufficient.

Indirect Cost Elements

An adverse social effect cannot always be replaced by an acceptable commitment of resources. Such an end is possible only if paying the cost of correction is more socially acceptable than suffering the social effects. Nonetheless, by assumptions (which are identified) of desirability of trade-offs, comparisons can be made, followed by additional analyses if necessary under other assumptions. The following paragraphs discuss how the policy analyst can quantify each cost element.

Where a particular set of alternatives require some *energy* resources, it should be, and to a significant degree must be, possible to estimate the energy requirements. This may be the equivalent in dollars per kilowatt-hour regardless how the energy is produced. For example, an effluent tax on waste water discharge from an industrial facility is likely to incur indirectly an energy cost. If the polluter wishes to avoid the tax because its cost is greater than the cost of installation of pollution control systems, he installs those systems which require a finite amount of energy to operate. Conversely, if the polluter chooses to pay the tax, then it is assumed that the community will use the money so derived to treat his waste for him. The addition of the waste treatment plant requires energy for its operation. In either case it is the energy cost that is the indirect effect of implementation of the policy alternative. For all these cost elements which are indirect effects of policy alternatives, we are talking about the absolute costs if they are implemented.

Time is defined here as potential productive labor time. It is essentially time that individuals must spend for unproductive purposes indirectly caused by the policy alternatives. If every person entering his plant must take a 15-minute physical examination, each person indirectly has lost a quarter hour from his basic reason for entering the plant. Similarly, if the absolute travel time to work or the variability in that travel time is changed, then the total number of opportunity hours is changed. Variability in travel time is important because if there are fixed work hours, then individuals must adjust their travel time such that they will not be late even when there is some unpredictability in travel time.

The value of time to be used in these evaluations may be made as complicated as one wishes. There is a large body of literature on the value of time and it is frequently an elastic measure. Frequently, the average wage rate may be used, especially if a set of alternative policy options are being compared. Obtaining an exact value is not necessarily important so long as the indirect effect of all the policy alternatives impacts the same population. In sum, time may be measured simply as the product of the number of hours, times the numbers of people whose opportunity time has been changed, times the average hourly wage rate.

Pollution is the indirect cost of air and water quality that would occur if a particular policy alternative were implemented. For our purposes the cost may be estimated in terms of current means available to obtain defined criterion levels of air and water quality. For example, if a particular policy option indirectly causes an increase in sulfur emissions in the atmosphere, it is possible to calculate the costs for additional technology needed to reduce those emissions to criteria levels. It should be noted that there may be a singularity that occurs in this process. If there is no available means, technological or otherwise, to reduce such pollution to the standard level, then by implication the cost must go to infinity. This is always the case if there is no ambiguity about the effects of the factor. This is especially obvious in areas impinging on public health. Finally, it should be clear that there are many policy alternatives that reduce pollution indirectly.

For example, a policy on transportation that would reduce the numbers of automobiles during peak hours reduces air pollution, at least carbon monoxide and oxides of nitrogen, and thus reduces the total cost of pollution control. Let us say that it costs $200 a vehicle to obtain pollution levels meeting socially acceptable criteria and that all vehicles will have to meet the standard. This means that for 100 million vehicles the total cost for pollution control will be $20 billion. If the number of automobiles needed for work travel is reduced by say 20 per cent, then the cost of reaching the pollution criterion levels will be $16 billion. Furthermore, if auto use is reduced by 20 per cent, it may imply that the design and operating precision of the control devices need not be as strict, which would mean still lower direct cost.

It should be noted that there may be a singularity in this costing process for pollution. If there is no means available, technological or otherwise, to reduce any increased pollution to criterion levels, then by implication the pollution cost must go to infinity.

For cost analysis purposes herein only two costs are included for *safety from accidental injury or death*. One is the direct costs of medical treatment; the second is loss of earnings. Although there are many variables in estimating the "true" total costs of accidents, these two components are generally accepted and in almost all cases provide a reasonable estimate within an order of magnitude of more sophisticated approaches. Furthermore, since the purpose is to *compare* alternatives, the commonality of the measure and the uniformity of application over all alternatives are important.

All alternatives that require some *material resources* effectively compete for other uses of the same commodities. Where the resource under consideration is renewable or in infinite supply, the use of natural resources has no external effect. It is only in the case of nonrenewable resources that the issue arises. It is not obvious how the cost should be calculated. Here the price at any instant in time does not always reflect the true costs of using up the resource. Petroleum products are a good example where the relations between price, use, and reserves are hardly in balance. It is likely that although the price will rise, it will not do so in relation to total supply.

To convert this indirect effect to dollars then, it is necessary to take into account the total limit on supply. We do this on the basis of a weighting of present unit cost by the ratio of the total supply to the total supply less the amount that the policy alternative requires for its implementation. That is,

$$C_T = C_P \left(\frac{S}{S - a}\right)^k$$

where C_T = the true cost of use
C_P = the present cost of use
S = the total resource supply
a = the amount required for the alternative
k = arbitrary exponent

If k is greater than unity, then the true cost of using a nonrenewable resource increases very rapidly as the denominator of the equation approaches zero. Practically, it represents a judgment on the value of the resource for social purposes other than the policy alternative under evaluation. For most important natural resources the exponent would be greater than unity, but rarely greater than 2.

In this analysis we are concerned with the external effects of nonrenewable natural resource use. The direct cost used in that part of the costing process is based strictly on present costs for the resource. The addition of an indirect cost is simply to take into account, in monetary terms, the future consequences of depletion of those resources.

Any policy alternative may change indirectly *health* conditions or the incidence of disease. A public policy calling for the chlorinating of all water supplies reduces the incidence of a whole series of waterborne bacteria. This in turn reduces the numbers of people suffering from associated diseases. Conversely, a policy protecting the sale of cigarettes increases the incidence of a whole series of degenerative diseases. To monetize these effects, the analyst can use an estimate of the cost of treatment for the disorders involved. This would include not only the direct costs, but also the indirect costs of lost time or earnings. Thus the public health costs are estimated in the same way as the accident costs.

Any policy alternative may require *land* for its implementation, land taken from other productive or socially useful purposes. Under such circumstances, takings may effectively change land use and in this way modify or degrade the local environment. This land-taking thus produces an externality which has an added value. For our purposes, it is the cost of providing an equivalent amount of land in the form that restores or retains the integrity of the use environment. Again, it is the dollar cost to obtain this equivalency that should be used.

Reliability can be considered as the losses in production due to the random failure of the alternative. All systems, social as well as technological, break down occasionally and cause stoppages in other processes. A power failure is an obvious example; a riot is another. In all such situations, these failures cost either time or production. Costs of both may be estimated in terms of either units of production and their associated costs or time losses to individuals. It should be noted once again that this is a very different cost from maintenance costs calculated under direct costs. These are costs that must be borne within the system to keep it operating or to return it to operation. The indirect reliability costs are those in which system failures incur penalties for other systems.

Job opportunity represents the concern with how a policy alternative may change the opportunities for jobs and the equivalent dollar benefits that may accrue from such a change in opportunity. In this case we are interested in the total number of job opportunities that are added or subtracted by implementation of the policy alternative. For example, a policy decision to rent cars for $1.00 a day to residents of cities, among other things, increases the numbers of people who would use a car to reach suburban jobs. Assuming perfection of match, then unemployed or underemployed people would be able to obtain full employment.

If the number of unemployed or underemployed in the region are known or may be estimated and the number of available jobs for them is known or may be estimated, then the total potential increment in income for these groups can be estimated. In terms of this evaluation procedure, the product of the number of jobs or the numbers of available people in the labor pool and the average wage rate define the income potential generated by the policy alternative. It should be noted that a policy alternative that increases job opportunity effectively reduces the cost of the alternative. Hence the entry under this cost element would be negative. Conversely, if a policy alternative reduced job opportunity, this would increase the cost of the alternative and the entry would have a positive sign.

Where a policy alternative indirectly changes the *shelter quality,* it has incurred a cost. In this case, the concern is wholly with degrading of housing.

If a policy alternative causes an industry to locate in a residential area, thus reducing the value and utility of the housing in the area, then the alternative has an attendant cost. The value of that change may be estimated from the cost required to provide the residents with housing of equivalent quality.

Whenever a policy alternative modifies the environment such that it degrades the *esthetic* quality of structures or environment, then that alternative has incurred a cost. The size of this cost is the replacement value if it is replaceable. If it is irreplaceable then the esthetic cost of the alternative goes to infinity. For the first case a finite monetary value may be estimated.

Crime is of concern in the analysis as a policy alternative may have an effect on social organization and change the probability of crime. In essence, where a policy alternative changes the environment socially or physically it may change the incidence of crime. It is the cost of this increase or decrease that forms the basis for monetization. If we assume that there is a direct relation between incidence of crime and the level of police services, we can estimate the costs of crime relative to the policy alternative. If the alternative increases the number of crimes, then an increment of police services is required to reduce the level to its original level. If the policy alternative decreases the frequency of crime, then the numbers of police are reduced. These costs are readily estimated and avoid a whole host of problems involved with estimating the cost of crime directly. Not infrequently changes in an area reduce crime in that area but lead to increases in other areas. It may be argued that the number of crimes committed does not change—only their locations. If that is the case, then crime is wholly independent of any policy variables and hence may be dropped from the list of externalities.

Application

There are three ways in which the results of cost effectiveness evaluations may be considered in policy analysis and development. One is the comparison of a set of policy alternatives. The second, is the derivation of savings between any set of policy alternatives and present system practice—status quo. The third is the comparison of subsystem changes within any societal system. Before discussing each of these uses or applications, certain characteristics of the model need to be described carefully.

The first such characteristic is the basic assumption that each criterion variable has a cost function that is determinate. In most cases, such functions are available and may be constructed. The major problem that is encountered is that the function is not linear over the whole range of application. In such a situation, the simple linear assumptions underlying the model do not hold, and serious errors in estimation may occur. Conversely, it is important to repeat that the purpose of this kind of analysis is not to provide exact costs or benefits, but to compare policy alternatives in a relatively crude way. As is often the case in policy evaluation, several alternatives fall within the same range of cost effectiveness. This fact, in and of itself, implies the requirement for a more detailed and sophisticated evaluation. Ultimately, of course, when a decision has been made on a particular alternative, such an evaluation is undertaken as part of the implementation.

A second characteristic of the model is that the pricing of the variables needs to be truly reflective of their value. If, for whatever reason, the unit cost of any element is much lower than its actual value in use, a serious error occurs in the evaluation. Land is a good example of a commodity whose price often bears little relation to true social value. If its price is based on its prior use rather than its potential use, then it is likely to be seriously undervalued. From a private enterprise standpoint this is precisely the path to profit. It is a poor basis, however, for public policy purposes. The same problem ensues for a variety of other elements, especially where the prices are artificial. This is the case most frequently in the areas of governmental regulation of prices such as natural gas, public utilities, and transportation. Frequently, these pricing policies are not economic in the sense of being reflections of market factors. Rather they are defined in terms of some class of public interest criteria.

A third characteristic of the model is the assumed constant relation between quantity and unit cost. That is, as the quantity of a needed commodity increases, its unit cost will not change appreciably. It should be clear that for a whole range of the evaluation elements this is not likely to hold indefinitely. This point has been made in relation to natural, nonrenewable resources, but it can also apply to other cost elements as well.

With these conditions in mind, the first way the procedure may be used is to sum the costs for each policy alternative for achieving a goal so they may be quantitatively compared (Table 3.8). The column totals are a measure of the total cost of the alternative to obtain a desired policy goal. The differences among the set of alternatives are a direct measure of the cost-effectiveness of each in relation to the rest. In effect, the method allows one to estimate the total cost differences among policy alternatives, and hence to determine the minimum cost members of the set. Only the total costs and costs differences should be employed for evaluation. That is, one should not attempt to take the ratios among the alternatives, for doing so would imply that the set of measures have mathematical properties that the method does not justify. In effect, it assumes that this costing procedure has interval properties but not ratio properties.

A second way in which the procedure may be used is to evaluate the row costs. It becomes possible to compare each alternative in terms of its relative costs on each element in the evaluation. Again the differences may be computed and analyzed. In many cases policy-makers implicitly and subjectively weight each cost element. Although this is not mathematically justifiable, it is a common practice. What the procedure forces, however, is that any such weighting be made explicit. This factor, in and of itself, can and should improve the quality of public debate over policy decisions. Furthermore, it should help separate the functional consequences of policy alternatives from the sociopolitical externality; for example, who shall control the policy that is implemented?

This evaluation procedure may also be used in relation to existing practice. Most frequently, alternatives are proposed to accomplish goals that are substitutes for existing methods. That is, for example, a proposal to use unit waste water treatment technology is proposed, because it is "better" than present central collection and treatment technology. If the present methods are included as an additional alternative in the evaluation procedure, their costs can be compared with the others. Again, this assumes that the total cost of present technology is based on the same dimensionality as the new alternatives. That is, if the concept of distributed

Table 3.8 Cost Effectiveness Evaluation Matrix

Resources Used	Candidate System				
	Population Redistribution	Cost Reduction Technology	Distributed Treatment Technology	Effluent Charges	Tax Incentives
Direct costs					
Energy					
Time					
Pollution					
Safety					
Natural resources					
Health					
Land					
Reliability					
Jobs					
Shelter					
Esthetic					
Crime					
Biotic					
Material					
Ethnic					

treatment is based on complete national application to obtain a desired water quality level, then the present system must be costed at the same national scale and to the same level of water quality criteria. If this is done, then the total column costs reflect the cost effectiveness of all the alternatives. Their difference from present technology is a direct dollar cost measure of the net benefit they would provide over existing methods. Thus differences among the row costs can be used as a direct measure of benefits or costs for any one alternative and the status quo or "do nothing" alternative. The sources of difference between them are explicitly described. Again, it is likely that the policy process will weight different elements; that is, energy use may be obscured if represented by the dollar cost.

The third application of the methodology is to effect a comparison among subsystem alternatives. Basically, the only difference between this use and the other already discussed is that many of the cost elements, especially the indirect ones, will have zero entries. This is the case simply because the larger system of which they are a part accounts for the major portion of costs in these areas. For example, if one wishes to compare alternative policies to reduce traffic accidents by say 50 per cent, this becomes a concern for a subpart of the highway transportation system. We might expect the first six externalities to be affected by a change in traffic safety, but not the last six, which are much more nearly related to the total highway transportation system design, location, and operation. In the end this makes the evaluation simpler and probably more accurate. But nonetheless, the basic approach is the same.

One thing, however, must be kept in mind with subsystem changes: Certain alternatives may differentially affect the larger system and hence change the overall costs or savings. Using the safety example, say one of the policy alternatives is a seat belt-starter interlock and another is electronic control of interchanges and intersections. The first has a set of direct costs and indirect costs for elements of safety, natural resources, and public health, but is not likely to impact any others. The electronic control devices have direct costs and impact more indirect cost elements because such a technology produces a change in traffic operations as well as safety. In cases like these, it turns out to be better for evaluation purposes to include the costs for the total system of which the policy alternative is a subsystem. In this way the policy analyst is less likely to miss a potential impact of the proposed alternative. Furthermore, doing the evaluation in this way, the relative effects of the subsystem proposals on total system operations are clearly observable.

The preceding discussion is based on the assumption of determinable costs for each criterion element. In fact, of course, this is rarely possible. We do not know exactly what the costs are, and we know they may vary over different locations or over time. These uncertainties may be quite significant in certain cases. Given this variability, one is faced with two alternatives. One is to define the appropriate distribution function and its parameters. Then the mean and variance can be used to estimate the true likelihood of the cost for that element. The second is to estimate the range of costs within which the true cost will lie. These two values can be used in the evaluation and thus provide an upper and lower boundary of costs. In essence, a minimum and maximum cost may be computed and a range of costs or cost differences derived for all the alternatives. This latter approach is simpler and probably more appropriate for most large scale policy evaluations.

This cost effectiveness method is basically different from that traditionally used in civil works evaluation, and is far closer in concept to the methodology devised for military weapons systems analysis. Essentially, the evaluation is a device for estimating the cost of a policy alternative required to meet a societal goal. The emphasis is on the expression, definition, and quantification in operational terms of that societal goal. This is in sharp contrast to the traditional practice in civil works evaluation of deriving an economic benefit-cost ratio.

The traditional benefit-cost analysis at its heart attempts to justify an isolated or given technology by demonstrating that the economic return from its use is greater than its direct cost. Its relation to operational societal goals can be quite indirect and frequently inscrutable. More importantly, the traditional evaluation methods have difficult mathematical problems whenever ratios of the external costs are calculated. As has been pointed out repeatedly, unless the individual scales of value can be shown to have ratio properties, the mathematical operations of multiplication and division are not permissible. It is highly unlikely that most of the scales used in social system evaluation have such properties. Consequently the benefit-cost ratios have no social meaning.

The evaluation methodology developed here starts from the assumption that explicit social goals can be and have been stated. If, and only if, this has been done, can the evaluation make any sense or serve any purpose. When such is done, the analysis is then capable of finding the bounds of the total costs to attain that societal goal. It is essential then that policy makers spend much of their effort in defining these social goals. In fact, it is only then that the analyst and the technologist can serve a useful social function. No policy analysis can proceed rationally unless the policy-maker has this goal judgment available. When it is available, the methods and procedure developed herein do permit an estimate of the costs of the alternative ways of achieving those goals, providing, however, that the cost includes the price for externalities as well as the direct costs. Thereafter, the policy-maker is in a position to make the necessary programmatic selection which is, in the end, the purpose of the policy analysis.

FEASIBILITY ANALYSIS FOR POLICY CHOICE

All other things being equal, the policy-maker selects that option that comes closest to satisfying the social goals at the lowest total cost *and* in the minimum time. Obviously, if rational and continuing analyses are an integral part of the policy process, then there should always be sufficient time for the selection and implementation of the optimal policy alternative to make it maximally cost effective.

Every physical or social technology has a finite life. Systems become obsolete, goals change, and new needs emerge. Thus there is a time-sensitive dynamic for all public as well as private policy. Beyond a certain point, the life of a policy alternative becomes too short to justify its costs, and its benefits will not be great enough to justify the social investment. There is a time window for every policy alternative and a primary objective of policy analysis is to determine the beginning and end point of that time interval. The objective of determining the temporal feasibility of policy alternatives is to provide a basis for estimating the appropriate time frame for their implementation.

Temporal Feasibility

Temporal feasibility of an innovation is basically composed of four distinct parts. The first is research and development time or the time required to create the alternative. The second is the engineering time to design and produce the alternative. Third is the operating time required to implement an operational system in specific locations. Fourth is the operational life span of the system, or the time within which the system performs its policy or goal function. The first three of these time components are related and effectively determine the start-up or implementation time. The life span is unique to each alternative and policy and must be evaluated separately.

One of the major failures of public policy in the United States has been the lack of recognition of the life-span characteristic of all alternatives. Consequently, the public policy-maker has consistently committed the country to social as well as technological systems that are presumed to be permanent and absolute means to attain social goals. Highway transportation is an example of a national commitment to one kind of mobility system as the ultimate solution to local person movement. It became evident that not only did that commitment produce enormous external costs, as well as benefits, but also the policy has been pursued to the point where the technology has become obsolete; other means are known which if developed would be far more efficient and economic local people movers for most urban purposes. If this had been recognized through a policy analyses in 1960, there would have been sufficient lead time to have created and laid plans for the implementation of the successor technology. In fact, one such program was initiated to provide such an analysis. The initiative was not successful and the result was another example of policy-makers being trapped through their own limited vision.

There are two ways in which the temporal feasibility analysis of policy alternatives may be carried out. One is to determine the time-cost function for each alternative. This is the way in which major space and military weapons systems programs are done. The second is to evaluate the relative costs of system alternatives over time. That is, the cost savings of any one or more alternatives relative to an existing system can be derived as a function of time. This usually produces a benefit function for the alternative. For example, if a personal rapid transit system is proposed as a substitute for highway transport in metropolitan regions, the cost difference for the same level of service between the two can be computed. The question simply is: What investment rates are required to attain the social goal in a fixed time period?

As an example of a time-cost feasibility analysis, let us evaluate three proposed alternative ways to reduce robberies: (*a*) a complete coverage television surveillance system with central station detection and dispatch; (*b*) a tripling of the number of police patrols; and (*c*) a brain wave detection system which transmits criminal intent. Let us also suppose that each of these alternatives will reduce robberies by 50 per cent which would save the country a dollar equivalent of $5 billion per year. Finally, let us suppose that alternative *a* would cost $15 billion; *b* would cost $10 billion; and *c* would cost $1 billion. These are the total direct costs for complete implementation as derived from previous evaluation. What we want now to know is how long it would take to achieve the desired crime reduction for these given direct costs.

Thus from the previous evaluation we know how much the direct costs of implementation would be for each of these three alternatives. We may now break these down into the seven direct cost elements. Let us assume that research and development is required in alternatives *c* and *a*, but not in alternative *b*. We assume that no installation is possible until needed research and development has reached a certain level of completion. Then it takes some time and dollars to produce the required technology for these two alternatives. However, to obtain the additional police labor for alternative *b*, a large training program has to be developed and initiated. This is the equivalent of a production cost. Finally, it takes time to get the system into operation across the whole country. Summing the cost in each of the categories of research and development, production, and installation, we can derive a cumulative time-cost function for each alternative. Each is plotted in Figure 3.2. As may be seen, with alternative *a* the social goal can be attained within a total of 10 years, alternative *b* within a total of 6 years, and alternative *c* within a total of 15 years.

If the goal were reached and the investment stopped at the end of the given time period, the analysis would be complete. The kind of function shown in the figure would be the basis for selecting the policy alternative. However, each of these alternatives assumes that in this case and in most policy situations there will be continuing costs. To maintain the achievement of the policy goal, that is, a 50 per cent reduction in crime, the direct costs of each alternative must be included with attendant operating costs. These, of course, include labor, replacement, maintenance, and overhead costs. In essence, we have only considered first costs. As is obvious, alternative *b* has a continuing cost approximately equal to its implementation costs, on an annual basis. In this example this is \$8 billion per 4 years or \$2 billion per year to which is added \$0.5 billion for maintenance and replacement. Thus the costs plotted over time continue to increase at the rate of \$2.5 billion per year.

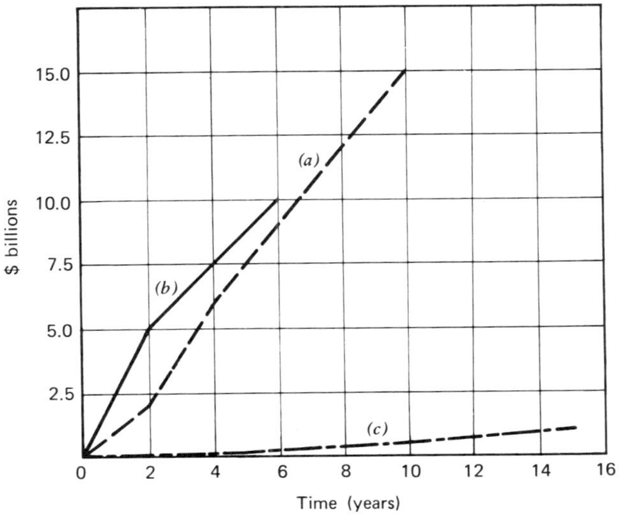

Figure 3.2. Time-cost function for three alternatives to achieve a policy goal. (*a*) TV surveillance; (*b*) police patrol; (*c*) brain wave detection.

For alternative *a*, the annual costs are only for replacement and for maintenance. If the life of each component element is 5 years, and the total system replacement costs are $6 billion, then the replacement costs would be $1.2 billion per year. If we arbitrarily assume that the maintenance costs will be $0.3 billion a year, then the time-cost function for this alternative continues to rise at the rate of $1.5 billion per year.

For alternative *c* the same problem exists as for the TV surveillance alternative. However, calculations assume that the cost of the system is only $0.1 billion per year for replacement. And that maintenance costs are very high, say $0.1 billion a year. Thus the total annual cost of obtaining a 50 per cent crime reduction with this technology would be $0.2 billion per year.

We can now plot the total costs of the alternatives over any arbitrary time period beyond implementation as shown in Figure 3.3. As may be seen, the total annual direct costs for each of the alternatives is quite different. The labor intensive police alternative is extremely costly over time to attain a given reduction in crime level. Alternative *c*, the brain wave detector, costs far less over time, but requires a wait of 5 or 6 years longer to reach the desired level of crime reduction. Because the costs for police patrol are higher than those for the TV surveillance, eventually the cumulative costs for that policy alternative exceed that of TV surveillance.

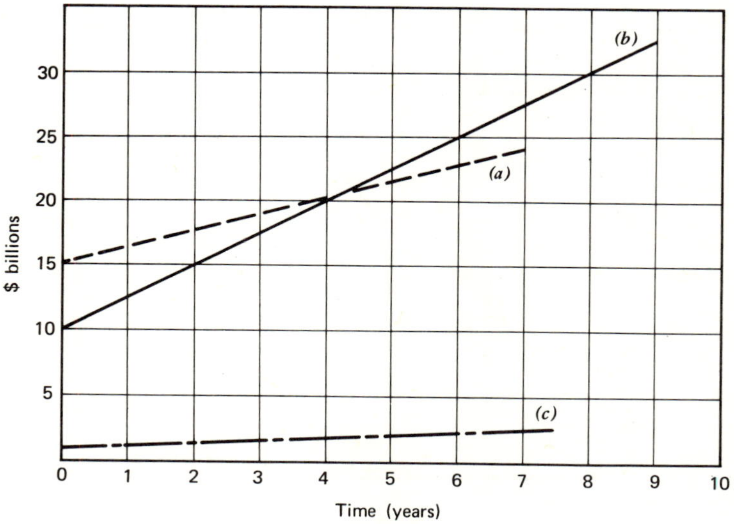

Figure 3.3. Cumulative annual variable costs of three policy alternatives. (*a*) TV surveillance; (*b*) police patrol; (*c*) brain wave detection.

Thus although the initial costs are less for police patrol, the ultimate annual costs make it a more costly alternative. We are, in effect, paying a higher long-term cost. Also, it is obvious that the brain wave detector alternative has a lower initial and cumulative costs. It just takes far longer to reach the desired goal than the other alternatives.

A simple time-cost analysis is not sufficient for a policy decision in cases such as this one. There is clearly a trade-off that has to be made between the time required to achieve the policy goal and the costs to achieve and maintain that goal. If we have two sets of simple linear equations, we can determine analytically the optimum alternative. Linear programming techniques are applicable and would provide a single valued solution. Where the equations are nonlinear but determinate, dynamic programming or similar operations research methods may be applicable. Under these circumstances a determinate optimum solution is also possible.

This discussion has been concerned wholly with the cost function. It is important also that we look at the benefit function. That is, for each dollar invested in a policy alternative, there is some increment in approach to the policy goal. In the present example, for every dollar invested in each of the crime control measures, a decrease in the number of crimes is expected. When each is completely installed and operating, a 50 per cent reduction in crime is expected. It is important, then, to evaluate this time related improvement in crime control. The simplest assumption is that there is a proportional reduction in crime that is directly and linearly related to the proportion of the policy alternative that has been installed and is operating. That is, the percentage of reduction in crime is directly proportional to the percentage of the system that is installed and operating at any particular time. In this case, the decrease in crime can be plotted over time for each alternative.

In the case of the TV surveillance alternative, we have estimated that it takes 7 years to install the system. Hence it will require 7 years to obtain a 50 per cent reduction in crime, or an average of 7 per cent decrease a year. For the police patrol alternative, it was estimated that it would take 4 years to implement. Thus the rate of crime reduction would be 12½ per cent a year. For the brain wave detector, it would take 5 years to obtain the 50 per cent reduction, or 10 per cent a year. The preceding averages refer to the period after implementation.

These time functions are shown in simplest form in Figure 3.4. From a policy standpoint, which alternative should be selected depends on how critical the problem is and how much resources are available. Under no circumstances can most policy goals be achieved immediately—by the stroke of the pen. Changing the social state takes resources including time. If the problem is critical, that is, time is short, then clearly the increased police

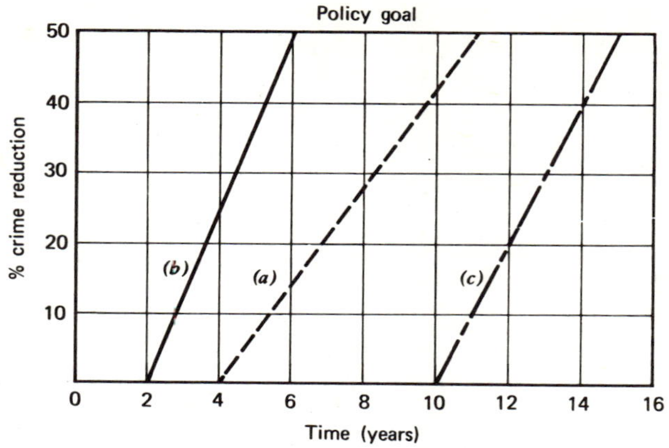

Figure 3.4. Cumulative rate of attainment of policy goals by three alternative means. (*a*) TV surveillance; (*b*) police patrol; (*c*) brain wave detection.

manpower alternative is the preferred choice even though it is the most costly. If resources are the most crucial consideration, then the brain wave detector alternative is the better choice. To conserve resources, time is sacrificed.

It is important to understand clearly the time-cost trade-off to achieve a policy goal. If the social issue is important enough so that time is the more important consideration, the policy-maker may select that alternative which provides a benefit most rapidly even if it is more costly. However, if the policy alternative to be implemented is a permanent one, then it is essential to examine the long-term costs. If one does not, then investments become cumulative and the society becomes burdened with high operating costs from the selection of policy alternatives that are not cost effective over the long run.

This short-term versus long-term benefit is a basic problem with almost all of the public works systems and most of the traditional social welfare systems. It may make sense over the short run, for example, to support conventional mass transit to serve those who cannot obtain an automobile. But this is an obsolete technology in most parts of the United States and the cost to keep it operating requires ever increasing public subsidies. This occurs along with its decreasing value to the regions which are supporting it. In essence, if the choice is to make investment decisions only for short-run gain, the risk is in causing long-run costs. This minimizes the probability of available resources to meet new or more important goals when they emerge or implement new methods for meeting existing goals which are more efficient and cost effective.

The analysis done to this point is simply one that estimates the time frame for attaining a policy goal. As much insight as this affords, it is really not complete. It does not take into account all the information available. Basically, we wish to provide the policy-maker with some estimate of the net benefit per dollar of resource investment. That is, we wish to provide some basis for indicating the cost effectiveness of each of the alternative means of achieving a policy goal. And we want to do this within a time framework. In the end we wish to show what the rate of return is for each alternative.

To do this we use our analysis of the net dollar benefits for each policy alternative. This is the end result of the analysis developed in a preceding section. The procedure produces dollar benefits or costs for all the direct and indirect variables. It produces a cost matrix of the form shown in Table 3.8. In the simplest case, all of the costs (the row headings in Table 3.8) would be zero except for two. One is the direct cost required to implement each alternative. The other is the one criterion function which it is desired to change. In the example we have used, this would be crime. Since the alternatives are expected to reduce crime, the dollar values in this column would be either a negative value, indicating a cost saving (a social benefit), or a positive value reflecting the total cost of crime in the country minus the saving from implementing the alternative. Since in our example, the decrease in crime costs is equal to 50 per cent of present costs, the dollar entry in this column would be the same for the three alternatives. We would have a 3×2 matrix as shown in Table 3.9.

The direct costs are those previously discussed. They would include all research and development, engineering, implementation, and operating costs. If time is not a consideration, then clearly the lowest direct cost alternative would be selected. However, if we wish to determine the rate of return to the society, we must take time into consideration. For example, if the total direct cost of developing, producing, installing, and operating the TV surveillance system over any desired time frame exceeds the savings in crime reduction, this alternative produces a net societal loss. The cost of

Table 3.9 Cost Analysis for Crime Reduction Policy

Policy Alternatives	Direct Costs	Indirect Costs
a. TV surveillance	x	k
b. Increased patrols	y	k
c. Brain wave detector	z	k

robberies is less than its cure! In the context of our policy analysis procedure, this would be an unacceptable alternative.

Conversely, if all the alternatives ultimately cost less than the crime costs, there will be a net benefit to the society. The question then becomes: When in time will the cost savings repay the social investment in the alternative? In the simplest case we can take each of the alternatives and determine their actual or prorated annual costs and then determine the cost savings they produce annually. The difference cumulated over time represents the net annualized return to the society. In the crime example this can be done simply. First, we determine the annual costs for each stage of the development, implementation, and operation of each alternative. Second, we determine the crime cost savings that accrue annually. Obviously, during the research and development and engineering phases, no savings accrue. Third, the costs are cumulated as are the savings and the difference calculated. Fourth, where these differences reach zero defines when in time the alternative produces a net benefit to the society.

Again let us use our crime control example. We assume that a 50 per cent reduction in robberies saves the country $5 billion annually. From the earlier analysis we can calculate the annual costs and savings. These are shown in Table 3.10 for each alternative. The computations are carried out to 20 years. The cumulated difference is simply defined by the equation:

$$Y = \sum_{i=1}^{t} (b_i - a_i)$$

As may be seen, the point in time where the net societal benefit begins to exceed the costs is different for each alternative. For the TV surveillance, this point occurs 11 years after the initiation of the investment. For the increased police patrol alternative, it occurs 6 years after initiation. For the brain wave detection alternative it occurs 14 years after initiation. We may plot this rate of return function as shown in Figure 3.5.

This function is a very useful one from a policy standpoint. It provides not only a measure of the time when the social investment is recovered by the society, but also the rate at which those returns accrue. Finally, this time function also indicates the long-term return from the investment in any of the alternatives. In our example this is quite clear. In the short term, increased police patrol provides the greatest return. However, its cumulative benefit grows more slowly than the other two. Consequently, after about 25 years the rate of return from the brain wave detector alternative exceeds both the increased policy patrol and the TV surveillance alternatives. Thus over the long term, this alternative is more cost effective. This exemplifies the importance in policy-making of distinguishing between long- and short-term feasibility. Should the society invest in the most cost

Table 3.10 Annual Costs and Savings Generated by Three Crime-Reducing Policy Alternatives[a]

	TV Surveillance Alternative			Police Patrol Alternative			Brain Wave Detector Alternative		
Time-Years	Cost *a*	Savings *b*	Cumulative Difference *c*	Cost *a*	Savings *b*	Cumulative Difference *c*	Cost *a*	Savings *b*	Cumulative Difference *c*
1	0.5	0	−0.5	1.5	0	−1.5	0.1	0	−0.1
2	0.5	0	−1.0	1.5	0	−3.0	0.1	0	−0.2
3	0.5	0	−1.5	3.0	1.7	−4.3	0.1	0	−0.3
4	1.0	0	−2.5	3.0	3.3	−4.0	0.1	0	−0.4
5	1.0	0	−3.5	3.0	5.0	−2.0	0.1	0	−0.5
6	3.0	1.0	−5.5	2.5	5.0	+0.5	0.1	0	−0.6
7	3.0	2.0	−6.5	2.5	5.0	+3.0	0.1	0	−0.7
8	3.0	3.0	−6.5	2.5	5.0	+5.5	0.1	0	−0.8
9	3.0	4.0	−5.5	2.5	5.0	+8.0	0.1	0	−0.9
10	3.0	5.0	−3.5	2.5	5.0	+10.5	0.1	0	−1.0
11	1.5	5.0	0	2.5	5.0	+13.0	1.0	0	−2.0
12	1.5	5.0	+3.5	2.5	5.0	+15.5	1.0	0	−3.0
13	1.5	5.0	+7.0	2.5	5.0	+19.0	0.2	1	−2.2
14	1.5	5.0	+10.5	2.5	5.0	+21.5	0.2	2	−0.4
15	1.5	5.0	+14.0	2.5	5.0	+24.0	0.2	3	+2.4
16	1.5	5.0	+17.5	2.5	5.0	+26.5	0.2	4	+6.2
17	1.5	5.0	+21.0	2.5	5.0	+29.0	0.2	5	+11.0
18	1.5	5.0	+24.5	2.5	5.0	+31.5	0.2	5	+15.8
19	1.5	5.0	+28.0	2.5	5.0	+34.0	0.2	5	+20.6
20	1.5	5.0	+31.5	2.5	5.0	+36.5	0.2	5	+25.4

[a] Note. (1) *c* is cumulative difference.
(2) Costs, savings, and differences are in billions of dollars.

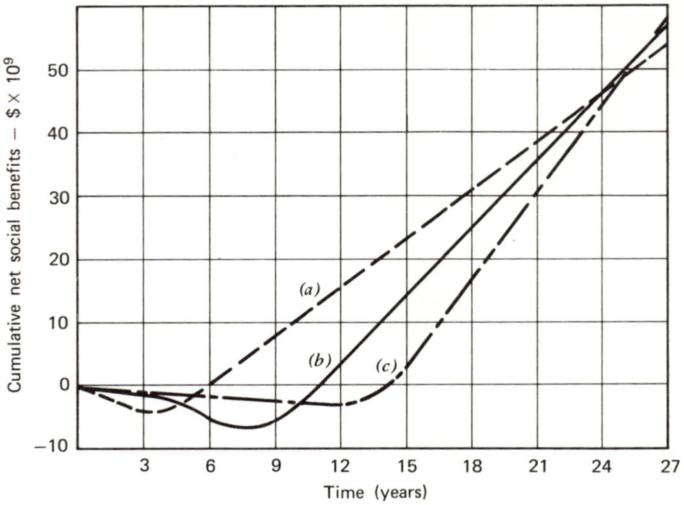

Figure 3.5. Time rate of return of three alternative crime reduction policies. (*a*) Increased police patrol; (*b*) TV surveillance; (*c*) brain wave detector.

effective alternative in the long term and wait 14 years for a net benefit? Or should it invest in the alternative that produces a net benefit in 6 years? Clearly, this is a political decision, not a technical one. However, the policy analyst must provide the basic information to the decision-maker so that such choices may be recognized and be made explicitly.

Finally, it should be pointed out that it is perfectly feasible to choose more than one alternative. In the present example, one could choose to implement increased police patrol and brain wave detection. At relatively low added cost, the society could obtain a net benefit (in about 6.5 years) from the first alternative as the second was being developed. As the brain wave detection system was being installed, the number of police patrols would be reduced and ultimately eliminated. The time rate of return can be determined for this case from the data in Table 3.10.

There is another approach to the temporal analysis of policy feasibility. It is concerned with the savings that may be obtained from a set of alternatives as they are related to a do-nothing alternative. This really becomes a question of the benefits of a set of alternatives over the status quo or present methods or systems of accomplishing a social end. Here, then, we are concerned with the rate of return from making a change in any societal system.

The process will be carried out using an example—the problem of providing highways in urban areas. In this case we can build the needed capacity with all of its attendant internal and external costs. Or we can eliminate most of the externalities by locating those freeways underground.[1] Let us further suppose, because of resource limitations, that we cannot make this transfer to underground space unless the time and dollar costs of subsurface construction is about equal to that on the surface. Thus a desirable policy goal is articulated, but the technical or economic capacity to implement it is not yet existent. The question then becomes: When and under what conditions could subsurface construction of highways meet the needs served by surface facilities?

Assume that the evaluation of savings from this alternative has been made by the procedure we have developed. Consequently, we know the equivalent total dollar savings accruing from the construction of freeways underground. We can then evaluate this saving as a function of time.

On this basis, if the benefits are plotted as a function of time (assuming equal construction costs), the two curves—surface and subsurface—have different slopes. On the basis of existing methods of economic analysis, the net benefit of urban freeways is approximately $2.5 million per mile per year (this excludes costs of indirect effects). From previous economic analysis, the incremental increase in monetary benefits can be calculated for urban freeways transferred to subsurface space. This increment is in the order of $0.8 million per mile per year.

Taking the case where there are no delays due to social or political resistance to the construction of freeways on the surface, the benefit function is simply a straight line function of time, as shown in Figure 3.6. If the policy decision were made to defer all construction until it could be located underground, how long would it be until the cumulative benefits equal that of the facility built on the surface without delay? The answer is determined largely by the time it would take to create a competitive underground construction technology (assuming research and development costs are negligible). During this phase, a loss of benefits to the society is recovered after construction begins. Since the slope of this function is greater than that for surface construction, the two curves eventually cross. This relationship is shown in Figure 3.6. If it is desired to have the total accrued benefits equal in a time period of say 10 years, then a competitive technology must be available in 1.33 years (+0.7, −0.3).

If the delay to surface construction due to social resistance is taken into account, a slightly different picture emerges. If the average construction delay for urban freeways is 2 years from the time of initial programming (a conservative estimate), there is a loss in benefits using surface construction because of the delay. As may be seen in Figure 3.7, the time available for

[1] The following two examples on underground construction are, in part, excerpted from Selected Reading 3.2.

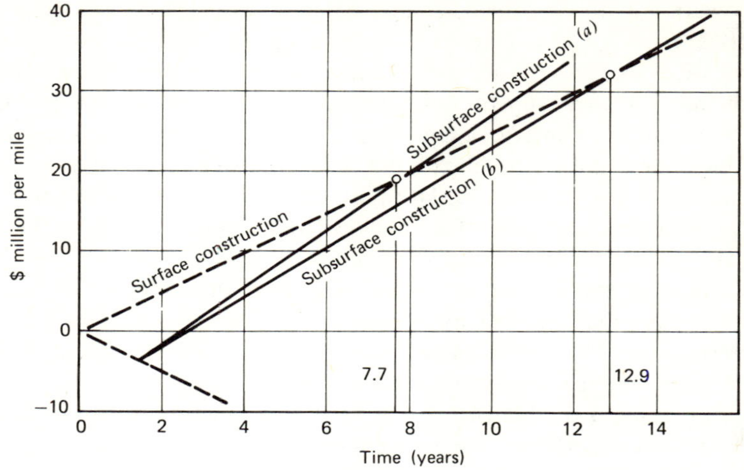

Figure 3.6. Economic benefits from urban freeways—no delay in construction. (*a*) Maximum estimated subsurface benefit; (*b*) minimum estimated subsurface benefit. *Source* Selected Reading 3.2, p. 54.

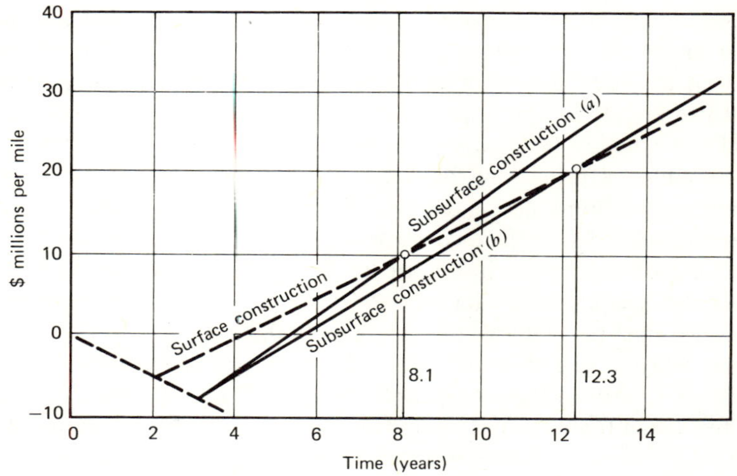

Figure 3.7. Economic benefits from urban freeways—two year delay in construction. (*a*) Maximum estimated subsurface benefit; (*b*) minimum estimated subsurface benefit. *Source.* Selected Reading 3.2, p. 56.

creation of a competitive underground excavation process is 3.1 (+0.4, −0.6) years. This is the time available if the total benefits accruing to the society are to be equivalent within a total time of 10 years. Obviously, any longer time period than 10 years may be taken as a criterion time interval and thereby provide a longer research and development and implementation period. However, intuitively, it would seem that the pay-off period for major public investments ought to be the order of a decade or less.

Under reasonable assumptions, a major national investment in transferring this one civil works function underground is highly sensitive to technological capability. If the state-of-the-art is within 2 to 5 years of producing a competitive underground excavation process, then urban highway needs can be met and, in being met, will produce a 30 per cent net increment in total economic benefits to the society. This can be accomplished within 10 $\pm$ 3 years. The issue from a policy standpoint then becomes twofold: (1) what is the likelihood of producing the needed capability within 2 to 5 years and (2) what institutional arrangements, including capital resources, are required to maximize that likelihood? In essence, this kind of analysis provides a means of establishing the requirements for a feasible technological alternative.

A further example may be meaningful: suppose a major policy decision requirement for funding the transfer of all urban electrical energy distribution systems to subsurface space. Assume further that it was estimated that the total national benefit from the transfer would be approximately $3 billion annually about 80 per cent of which would accrue to metropolitan areas. Given the fact that about one-third of all distribution miles are already underground, there is a total net benefit potential of approximately $2 billion per year. If a public investment to make the transfer is proposed, then there are, at the simplest, three factors to consider in evaluating that investment: (1) the costs, including those required to pay off surface facilities with remaining life; (2) the rate at which the transfer may be made; and (3) the rate at which the added benefits accrue.

To dimension the problem, assume that there are approximately 667,000 miles of electrical distribution lines on the surface in the 50 largest urban areas. Using a representative cost of $8000 per mile to install those lines (and hence the cost requirement for locating them underground), the total cost of the transfer is approximately $5.3 billion. (Costs are on the order of $36,000 per mile for the current technology for undergrounding distribution lines.) Doing this would provide an incremental benefit of the order of $2 billion or $3000 per mile of utility lines in urban areas.

Since the capability does not exist to make this transfer at this cost, the total cost must then include research and development costs and a deferral of benefits. If it is assumed that it takes $50 million per year and 3 years to produce a competitive technology, then it is obvious that it requires a total of approximately 8 years to recover the total investment costs assuming a 10-year period for replacement of all surface distribution systems after the capability has been developed. The net benefit function is shown in Figure 3.8.

In the preceding two examples we have simplified our policy decision by insisting that the alternative have the same unit cost as the present method. This is an extremely useful device to provide a comparison of the benefit functions for the alternatives. We could have, of course, examined several additional policy alternatives and made a multiple evaluation.

There is no requirement and certainly no necessity for making the assumption of cost equivalency. If the benefit function is essentially a positive one for any alternative, then we can compare the dollar equivalent of the relative benefits with the dollar costs of creating, producing, implementing, and operating the alternatives relative to present methods. The costs can be plotted against the benefits, and as in the previous two examples, one can find the minimum cost differences between the alternatives. In

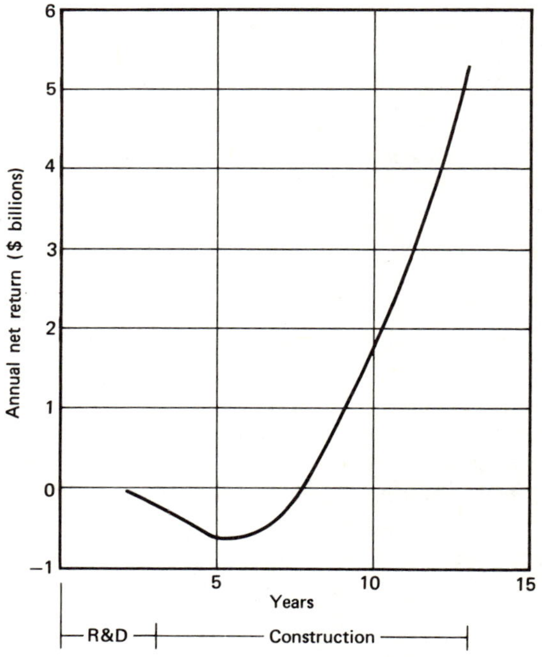

Figure 3.8. Expected return rate from undergrounding urban electrical distribution lines. *Source.* Selected Reading 3.2, p. 57.

the end this is a comparison of the benefit/cost ratios among the alternatives plotted over time. In this fashion, the trade-off function can be estimated.

Total Feasibility Analysis

The evaluation procedures that have been described have been developed on the assumption that policy feasibility can be indicated by the criteria of time and costs. Clearly, these are necessary but not sufficient. Also, there is considerable uncertainty in the estimates. In fact, the whole policy process is filled with uncertainty. Since we do not have any real knowledge of the magnitude of that uncertainty, we are in the position of having quantitative estimates of uncertain processes. This is not an unusual problem in policy analysis, but what is needed is a way to recognize that uncertainty, and build it into the evaluation. The fact is that we normally are unsure whether it is possible to create in real operational terms the kinds of alternatives that can be conceived. Scientists and engineers can look at a system concept and get a feeling for its feasibility. The fact that this human judgmental process goes on, gives a clue as to how to evaluate the more uncertain elements of feasibility.

Rating scale techniques again present themselves as a first-order means to quantify the professional judgment of feasibility. However, before developing the rating scales, it is important to define the domains in which judgmental measures are essential. When there are quantitative and objective means to evaluate alternatives, they should be used. We are concerned here with the five dimensions or areas where expert judgment is beneficial.

TECHNOLOGICAL FEASIBILITY

Within physical limitations of the natural universe, any system is potentially possible. The question is: Can such a system actually be created and operated? At the outset, the system concept as a possibility must be distinguished from a feasible system. Thus technical judgment is desirable concerning how easy or difficult it may be to translate a concept into a working system. This kind of judgment must necessarily come from experts and specialists in the area. Basically, the question we wish to ask them is: How difficult will it be technologically to develop this alternative?

SOCIAL FEASIBILITY

The concern here is with social acceptability of any policy alternative. For policy analysis purposes, the public official, especially at the local level with local issues, is a sensitive judge of the social feasibility of any public policy alternative. The question to be answered can be stated: How socially acceptable is the policy alternative?

ECONOMIC FEASIBILITY

Basically, the issue concerns the magnitude and acceptability of the magnitude of economic resource requirements. The question is whether the resource requirements for a particular policy alternative exceed limits of economic prudence or economic capability. Again this is largely an expert judgmental problem of the form: How acceptable are the resource requirements for this alternative?

POLITICAL FEASIBILITY

Here the issue is whether any particular policy alternative receives initial and continuing support if an attempt is made to implement it. This is a judgment concerning the impact of the alternative on the stability and continuity of political institutions. If it tends to upset the balance of forces within the society or threaten the power base of political incumbents, then it is not a feasible alternative within the context of the political organization structure and will be rejected. The question simply is: How politically acceptable is the alternative?

TEMPORAL FEASIBILITY

A policy option that cannot be implemented in a reasonable period of time is not socially or politically acceptable. For almost all policy decisions there is an expectation of results accruing within a defined time period. From the temporal analysis of policy alternatives an estimate of the time frame for a policy's implementation is available. The decision-maker must select that alternative which, among other things, provides the social benefits within an acceptable time. The judgmental question for expert reaction is whether the time frame required for implementation is acceptable.

SUMMARY

There are thus five dimensions of feasibility for which expert judgment is desirable before a final decision on a policy alternative can be made. To achieve this judgment a simple rating scale can be used for each (see Figure 3.9). These scales should be rated by qualified professionals and experts.

Assuming reasonable objectivity on the part of the rating groups, a reliable measure of feasibility can be obtained. Each of the items can be

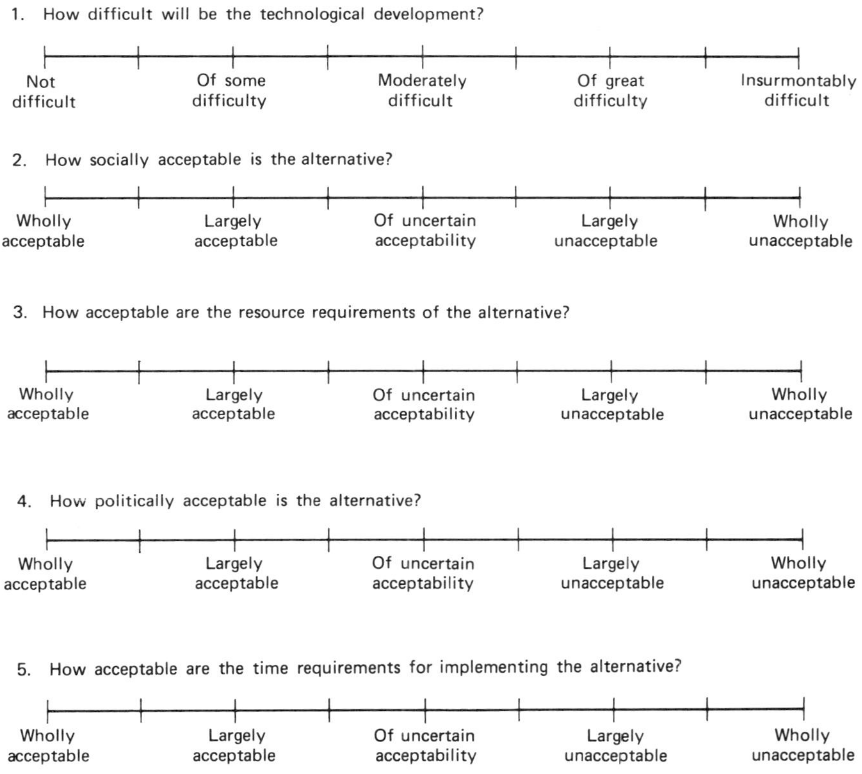

Figure 3.9. Measures of feasibility.

treated as independent attributes in which the units on the scale are true interval values. As with other such items the mean ratings may be summed to obtain an overall feasibility score. Generally this is a weak assumption to make for scaling, but for policy analysis purposes it is useful. At least it is more systematic than the chaotic practice for public policy-making prevalent in the early 1970s.

SELECTED READINGS

3.1 *The Analysis and Evaluation of Public Expenditures: The PPB System*, Joint Economic Committee, Congress of the United States, 91st Congress, 1st Session, Vol. 3, 1969.

3.2 Baker, R. F., R. M. Michaels, et al., *The Use of Underground Space to Achieve National Goals*, Special Report, American Society of Civil Engineers, December 1972.

3.3 Coates, V. T., *Technology and Public Policy*, George Washington University, 1972 (PE-211-453).

3.4 Edwards, A. L., *Techniques of Attitude Scale Construction*, New York: Appleton Century Croft, 1957.

3.5 Forrester, J. W., *Principles of Systems*, Second Preliminary Edition, 1968.

3.6 Forrester, J. W., *Urban Dynamics*, Cambridge: The MIT Press, 1969.

3.7 Foster, B. E. (Ed.), *Performance Concepts in Buildings*, Vol. 1, National Bureau of Standards, 1972 (COM-72-10309).

3.8 Gilmore, J. S., *Environmental Policy Analysis, Parts I and II*, Denver Research Institute, August 1971 (PB-206-875).

3.9 Guilford, J. P., *Psychometric Methods*, New York: McGraw-Hill, 1954.

3.10 Jones, M. V., *A Technology Assessment Methodology*, The Mitre Corporation, Washington, D.C., 1971.

3.11 *Public Transportation in the Chicago Region—Present Performance and Future Potential*, The Transportation Center, Northwestern University, 1973.

3.12 Romney, A. K., R. N., Shepard, and S. B. Nerlove, (Eds.), *Multidimensional Scaling*, New York: Seminar, 1972.

3.13 *A Study of Technology Assessment*, Report of the Committee on Public Engineering Policy, National Academy of Engineering, Committee on Science and Astronautics, 1969.

3.14 *Technology: Processes of Assessment and Choice*, Report of the National Academy of Sciences, Committee on Science and Astronautics, U.S. House of Representatives, 1969.

3.15 Torgerson, W. S., *Theory and Methods of Scaling*, New York: Wiley, 1958.

PART 2
CASE STUDIES

By the early 1970s, many of society's problems were common knowledge, but a systematic, rational approach to policy development was not available. Part of the reason was the absence of data and theory for predicting the social effects of alternate actions. But most of all, an elitist perspective prevented operating professionals of many types from the collaboration that is fundamental to rational policy development.

This elitist concept of political thought was a natural extension of man's fight for survival—the feeling that one's views of the universe and the day-to-day struggles were "right." As a corollary, of course, those who did not agree were the "enemy" and were "wrong." Social systems had become more complex with increased population; more sophisticated technology had shrunk distances by improved communications and transportation; and agriculture, industrial, and medical advances had removed many routine worries. During the period of scientific growth, "compartments" had become so distinct that fields of specialization became sacrosanct and were viewed as the ultimate authority, particularly by the elite within a given area of specialized knowledge.

Certainly in the early part of the twentieth century the engineers, scientists, and technologists with the capability of accomplishing a task that required specialized knowledge were quite influential in determining both *whether* the task should be attempted and *how* it should be done. In Chapter 1, this has been referred to as the tyranny of science, technology, and engineering. The interrelationships were poorly understood however. Particularly in public works, the disconnection between societal functions prevented professionals from obtaining a comprehensive perspective of alternative solutions to community problems.

And so even into the 1970s, there were editorials in engineering and scientific journals on "professional versus political" and on the loss of prestige of scientists and engineers. The pursuit of the former powers to influence policy as in the "good old days" was thinly disguised by pleas that only the profession has enough know-how to decide on such matters. But the backward looking viewpoint represented in such editorials all too frequently failed to grapple with the essence of the problem. Unless one first studies and understands *the needs and desires of individuals* who are affected by a public work, and the political processes by means of which these needs and desires can be most effectively provided, public works continue to be undertaken which are unnecessary and/or which cause nearly insoluble problems after construction. If the *policy* is not well thought out, how can its *results* be good?

The following chapters provide examples of how problems in policy development can be approached, particularly if existing political systems can be made more responsive to rational analyses of alternatives. During the final years of the twentieth century, policy development will become much more susceptible to objective analysis, but the changes will not appear to come quickly. Furthermore, the initiative for the change can come most easily from engineers and other operating professionals who understand the technical complexities of one or more components of society. If these professionals do not furnish the new approach, then a new discipline will have to evolve, one that can comprehend the complexities of the components and can also comprehend the needs of the social system.

The cases that are presented in Chapters 4, 5 and 6 are typical of the "muddling through" process which characterize the initial efforts in policy development. None of the examples follow closely the suggested procedures of Chapter 2 and 3. Rather, some of the most simple applications of policy development are presented.

Recall that policy development has been defined as a process designed to improve the allocation of resources. No exclusions have been as yet identified. Until such exclusions have been made, each decision that implicitly or explicitly decides how resources are expended is in the context of this book a policy decision. While the focus has been and will continue to be on the public as opposed to the private sector, this exclusion was not on the basis of principles involved but to simplify and limit the discussion.

Each of the case studies then is a valid example of policy development. Each is concerned with the question of applying new principles that provide for problems of resource allocation a better link between the technical and political processes.

CHAPTER **4**

POLICY DEVELOPMENT FOR AN ONGOING PROGRAM—HIGHWAY SAFETY

The problems related to improving highway safety are particularly adaptable to systematic and objective policy analyses. Because of oversimplified views as to how corrective actions should be made, and because of crusading interests for various solutions, the problem is also typical of why rational decision-making by the political system was difficult in the 1960s.

In the following chapter, the problem of resource allocation is examined for the conditions that existed in the early 1970s. These conditions were dynamic and were changing from year to year. The importance of this case study is not in the results that are developed, but in the thought process and perspectives that are indicated.

Three types of problems in resource allocation are examined: (1) analyses for establishing goals; (2) evaluation of an existing policy; and (3) policies for local programs. Each of these problems require policy decisions; that is, resources must be allocated or not allocated.

The case study that follows is typical of policy development from the perspective of those responsible for or concerned with improving an ongoing program. The thrust of such analyses is to increase the effectiveness of a program. Because of ultimate impacts of the program on the total social system, the analyses must include the total consequences of the resource allocation.

The example is also functionally oriented in that totally new, innovative ways to accomplish the basic purposes of a program are not analyzed. The focus of the analysis is on improvement rather than replacement. The latter problem is one that is conducted by a higher level organization whose responsibilities transcend those for a specific program. The following case study is nonetheless an example of policy development—analyses for the guidance of policy-makers.*

* The manuscript for Chapter 4 was in final form prior to the energy crisis of 1974, the imposition of the 55 mph speed limit, and the unexpectedly large reduction in highway deaths. These events provide an unusual opportunity to analyze how reductions in highway fatalities can be accomplished. Chapter 4 is primarily addressed to goal setting rather than alternative solutions. Nonetheless, the changes in 1974 provide an additional dimension for the reader. Suffice to say, by the end of 1974 there was a lively debate in progress relative to what produced the reduction in fatalities—whether it was due to the speed limit or to the impact of fuel shortage on driver attitude, types of drivers (less use by younger drivers), less vehicle-miles of travel, and similar factors.

NATIONAL POLICY MAKING

Highway safety as an example has both good and bad features. The perception of need is well documented, but the abundance of individual solutions creates serious communication problems; that is, it is difficult to divorce a given analyst from his preconceived notions. Highway safety is also a very complex problem to which whole books have been addressed. To compress these ideas into a few pages requires an oversimplification, and to circumvent this difficulty, the following pages make liberal use of earlier works.[1]

On the other hand, the consequences of not having explicit policy goals—a not unusual condition for public programs—makes the need for such goals much more apparent. Even though the application is for highway safety, a parallel approach for other programs is not difficult to envision.

To begin, the example makes several assumptions. (1) The hypothetical study begins in mid-1973, and deals with conditions as they actually existed at that time. (2) The program to be initiated as the result of the study is to begin on or about January 1, 1974. (3) A report on the study itself must be turned over to the policy-makers within 3 months after the study is initiated by the engineers. (4) The study is being developed for a joint committee of the Congress which is composed of committees responsible for the Federal Highway Safety Acts. (5) Data are available from many sources, but not in the form desired, and need to be converted to become usable.

The relatively short time frame was chosen as being typical of the amount of time usually available for the development of national policies, if not longer. Furthermore, in real life, time to develop legislative action must also be taken into account as a variable. The availability of data from many sources is also a typical condition, as is the need to convert the data before they can be used. These time frames suggest the importance of having data and analyses already performed, or of having a continuing process of policy evaluation.

The following steps in policy development are available from Figures 2.1 and 2.4:

- Perception of need
- Goal definition

[1] See Selected Readings 4.2, 4.3, and 4.4.

Perhaps the most significant fact was that safer conditions were produced, with a loss of mobility but with a savings in resources. The overall economic impact was still in question. There is also a possibility that a change in value system had occurred (see Chapter 3, Table 3.2). The unpredicted events of 1974 emphasize the importance of estimating the effects of a change in one function (energy provision) on other functions (transportation) and various criteria (life expectancy).

- Policy analysis
- Alternative selection
- Resource allocation

These are the steps in the analyses that follow.

Perception of Need

The major issue in highway safety is that the fatalities, injuries, and economic losses from highway accidents are unnecessarily large. Beginning in the 1920s, periodic references to the horrors of automobile accidents became common. Ever since, over holiday weekends, grim predictions followed by the announcement of the results have provided a kind of morbid box score.

The statistics have been continuously discouraging. Data such as those summarized in Figure 4.1 are provided annually by the National Safety Council. Pleading, cajoling, and cooperative efforts between government and industry have not been sufficient to stop the steady increase in annual numbers of fatalities. By the early 1970s, the total number killed on United States highways had approached 2 million.

Although traffic accidents are universally deplored, there are different opinions on which of the consequences are the worst. Fatalities are not the only concern: injuries and economic losses are also cited. Somewhat defensively, the highway interests have noted that there has been a steady decrease in the *rate of fatalities* (deaths per mile of vehicle travel). Unfortunately, the increase in travel more than offset the decrease in rate, and the total loss figures steadily mounted.

The perception of need was expressed by the highly effective crusaders—a group of individuals who felt particularly strongly about the condition. But in spite of headlines, editorials, and oratory, highway safety was not a common topic of conversation in the United States—until a relative or close friend became a casualty.

Certainly an observant individual would have good reason to voice a complaint about highway safety, particularly since the problem appears to the layman as one that can be relatively easily solved. Certainly the political system should be expected to examine the problem quite critically.

Goal Definition

Given a vocal perception of a need, the political goal is to fulfill that need, or to quiet the demands for action. The political goal for highway safety,

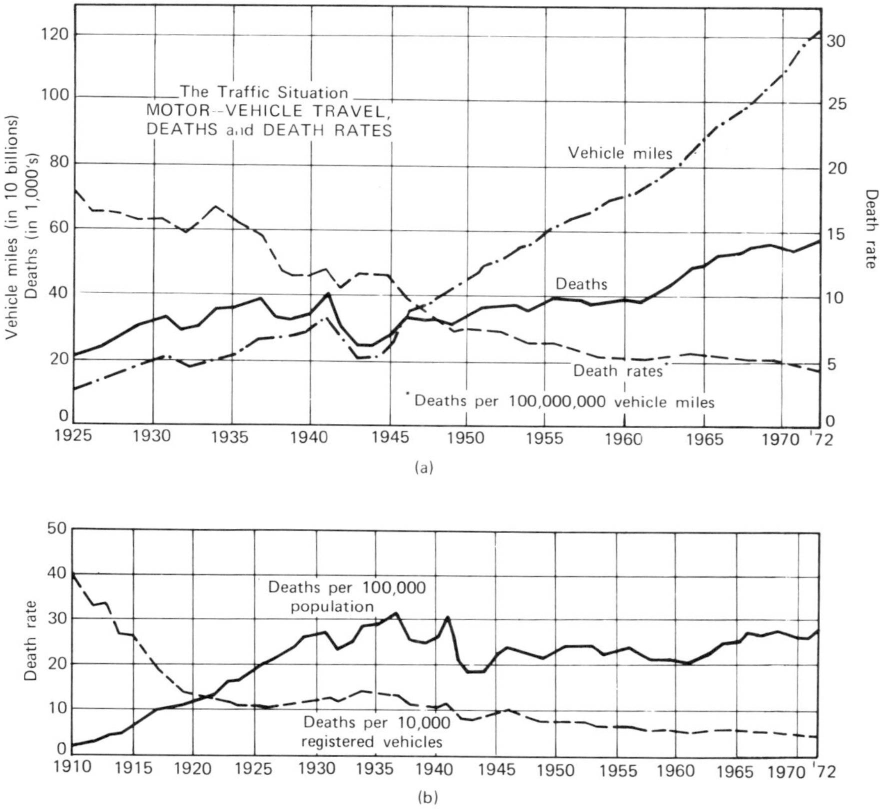

Figure 4.1. Growth and other changes in highway transportation and traffic safety. *Source:* Selected Reading 4.1, pp. 40–41.

then, is simple to state—to stop highway accidents—eliminate the "unnecessary loss of life." Political goals are usually relatively easy to state, particularly when the goal cannot provoke opposition.

Political goals do not become *policy goals,* however, until some action is taken. To go from a political goal (or perceived need) to effective policy action, specific policy goals are needed, which require answers to the following questions:

- What is the relative severity of the condition?
- What is the range of feasible goals?
- Are there constraints?
- How should the goal be stated?

In defining the goals, consideration of alternative means to achieve the objective is of minor concern. The selection of a solution requires detailed analyses of alternatives, but such studies are not particularly relevant to the question of deciding whether something *should* be done. It is sufficient during goal definition to establish that the goal is technically feasible within a desired set of constraints on resources, time, and any other relevant social, economic, environmental, or political consideration.

RELATIVE SEVERITY OF THE PROBLEM

Death, injury, and economic losses are the consequences of highway accidents. Of these, death and injury are related to a very basic aim of each individual—a long life and the fewest possible days with disability from sickness and/or injury. Economic loss is easily catalogued, and can readily be compared with expenditures necessary to reduce the losses.

A comparison of the major causes of death in the United States in the early 1970s is provided in Table 4.1. From this type of data, one can see that highway accidents are a significant factor is shortening the lives of some individuals—approximately 3 per cent of the deaths each year are from highway accidents. While it is true that cancer and heart disease

Table 4.1 Major Causes of Death in 1967[a]

Causes	Total Number	Rate[b]	Risk[c]
All causes	1,851,323	936	107–1
Heart disease	721,268	365	365–1
Cancer	310,983	157	640–1
Stroke[d]	202,184	102	980–1
All accidents[e]	113,169	57	1,750–1
Pneumonia	57,636	29	3,450–1
General arteriosclerosis	37,564	19	5,250–1
Diabetes mellitus	35,049	18	5,550–1

Source. From Selected Reading 4.3, p. 24.

[a] Based on *Accident Facts*, National Safety Council, 1969 Edition, p. 8 and 9.

[b] Deaths per 100,000 population.

[c] Average chances against death at start of 1967, varies with causes and age. For example: infant mortality, 44 : 1; heart disease (age 45 to 64), 226 : 1; and heart disease (age 25 to 44), 2,400 : 1.

[d] Vascular lesions affecting central nervous system.

[e] Motor vehicle accident figures: total, 52,924; rate, 27; and risk, 3,800 : 1.

produce a much greater toll, the number of highway accident fatalities is as great as those from all other accidents combined.

To obtain additional insight on the relative severity of highway fatalities, the risk of death can be examined, where risk is defined as the mathematical probability of an event. Risk can be estimated by a conversion of population rates. Thus if there are 50 fatalities per 100,000 population, then the risk is 50:100,000 or 1:2000. Probable risks are best developed with disaggregated statistics, that is, by placing individuals with comparable groups. Nonetheless the gross statistics of Table 4.1 provide a useful perspective. Thus one of every 107 individuals, selected at random, will die from some cause each year. For heart disease, the effect of grouping can be seen in Table 4.2, wherein for the total population, one of every 365 will die from this cause, while for the age bracket of 45 to 64, one of every 226 will be a casualty of heart disease.

For highway safety, accidents are less a function of age, although Table 4.2 shows that the greatest risk (1:2000) is for the age 15 to 24, as compared to 1:3800 for all ages and 1:10,000 for the age groups under 15 years. These rates do not account for any differences that exist in the amount of travel.

To determine the relative severity of highway safety as a cause of death, analyses such as those summarized in Tables 4.2 and 4.3 can be conducted. From these tables, the following observations can be made:

- Highway safety is *not* the largest cause of death for any age group.
- The chances of being a highway fatality are much more remote than are the chances of dying from heart disease or cancer for ages greater than 44.
- The risk from highway accidents is quite comparable to risks from all other types of accidents combined.
- Accidents of all types, and particularly highway accidents, are the greatest risk for the 15 to 24 age group.
- The risk of death from "other causes" for the 0 to 4 age group is much greater than is the risk for highway accidents for any age group.
- If the risks from heart disease were to be reduced to a risk comparable to that which existed for highway accidents, the annual savings on life would be approximately 700,000.

Tables 4.2 and 4.3 indicate that risks of death are much greater for several categories than for highway accidents. On the other hand, the risks are significant, and involve quite different types of phenomena; that is, heart disease and cancer represent physiological problems, whereas

Table 4.2 Risk of Death by Various Causes—Early 1970s[a]

Age Group	Population in 1969 (1,000)	All Causes: Total Deaths	All Causes: Rate per 100,000	All Causes: Risk 1:No.	Highway Accidents: Deaths	Highway Accidents: Rate per 100,000	Highway Accidents: Risk 1:No.	Other Accidents: Deaths	Other Accidents: Rate per 100,000	Other Accidents: Risk 1:No.
0–4	17,960	89,301	2,312	43[b]	1,987	10	10,000	5,276	33	3,040
5–14	41,345	17,542	43	2,300	4,105	10	10,000	4,264	10	10,000
15–24	35,050	41,140	124	810	16,543	50	2,000	6,469	18	555
25–44	47,990	112,679	239	418	13,602	29	3,450	10,082	20	5,000
45–64	41,400	479,038	1,174	85	11,031	27	3,800	12,865	31	3,220
65–74	11,955	453,546	3,849	26	4,261	36	2,280	6,700	56	1,780
75 and over	7,515	736,836	10,056	10	3,333	45	2,220	7,096	95	1,050
Total	203,216	1,930,082	966	104	54,862	27	3,800	60,002	30	3,330

Age Group	Heart Disease			Cancer			All Other Causes		
	Deaths	Rate per 100,000	Risk 1:No.	Deaths	Rate per 100,000	Risk 1:No.	Deaths	Rate per 100,000	Risk 1:No.
0–4	—	—	—	—	—	—	83,225	460	218
5–14	—	—	—	2,589	6	16,670	6,584	16	625
15–24	—	—	—	2,731	8	12,375	15,396	44	2,280
25–44	19,521	41	2,440	18,492	39	2,800	50,982	96	1,040
45–64	181,446	445	225	116,385	285	385	157,311	37	2,700
65–74	198,856	1,687	59	91,240	774	130	252,489	1,296	71
75 and over	342,823	4,679	21	85,775	1,171	85	290,559	4,066	25
Total	744,658	373	268	318,547	159	630	752,013	380	260

[a] Based upon Selected Reading 4.1, p. 8. Highway and other accident data are for 1970; other data are for 1968.
[b] Rate for 0–1 equals 2,226; risk equals 45. Rate for 1–4 equals 86; risk equals 116.

Table 4.3 Comparable Risk of Death for Age Group and Causes[a]

Risk Group (1:No.)	Age Group	Cause of Death
20–100	75+	Heart disease, cancer, other causes[b]
	65–74	Heart disease, other causes
100–1000	65–74	Cancer
	45–64	Heart disease, cancer
	15–24	Other accidents
	5–14	Other causes
	0–4	Other causes
1000–3000	75+	Other accidents, highway accidents
	65–74	Other accidents, highway accidents
	45–64	Other causes
	25–44	Other causes, heart disease, cancer
	15–24	Highway accidents
3000–5000	45–64	Other accidents, highway accidents
	25–44	Other accidents, highway accidents
	0–4	Other accidents
+5000	15–24	Cancer
	5–14	Other accidents, highway accidents, cancer
	0–4	Highway accidents
No record	15–24	Heart disease
	5–14	Heart disease
	0–4	Heart disease, cancer

[a] Based on data in Table 4.2.
[b] Other causes, see Table 4.2.

highway and other accidents appear to involve primarily physical and psychological factors. Furthermore, highway accidents involve a *hardware system* failure as compared to a failure in *human systems* for heart disease and cancer.

On purely rational grounds, the greatest potential for improvement in life expectancy lies in reducing deaths from heart disease and cancer. On the other hand, there is great emotional reaction of individuals to catastrophes or to what appear to be "unnecessary" causes. Also, the failure of a physical, man-made system (crash of an airliner) is not as "acceptable" as are catastrophes from natural phenomena (earthquakes and floods).

Accordingly, even though a reduction of highway fatalities does not offer the same potential for increasing life expectancy as do some other reductions, significant risks do exist, and possibilities for improvement appear to be well within reach. The effectiveness of the solutions for improving highway safety, however, should be compared with the effectiveness of other means for increasing life expectancy.

The other losses from highway accidents—injuries and economic losses—are of less consequence to individuals than is death. Improvements in these conditions can be considered as either secondary or primary goals. Should either become primary benefits, the social goal is different (than if reduction in fatalities is the primary goal), and the effectiveness of the solution needs to be compared with other methods for achieving the same benefits.

The question of reducing the *rate* of fatalities per mile of travel is a secondary question insofar as social goals are concerned. For government and industry administrators within the highway system, however, the objective of reducing the fatality rate is always present, even though it is not the primary determinant. For functional goals, the relative effectiveness of solutions is based on reducing the risk per mile of travel rather than the total losses.

RANGE OF FEASIBLE GOALS

By the early 1970s, there was a 50-year history of the consequences of highway accidents. Figure 4.1 shows the growth in total losses and in rates since 1925. Analyses show a steady growth in annual fatalities, which means that unless effective action is taken, the annual number of fatalities is likely to continue to increase. In fact, the amount of the *increase* per year is also likely to get larger.

The increase in accidents has been approximately proportional to the increase in travel. Accordingly, a reduction in accidents (and fatalities) can be expected as a result of a reduction in travel—either in the growth patterns or in total travel.

The relation between risk, mobility, and resources is described in detail elsewhere.[2] Thus for a given highway system, mobility, risk, and resources are inseparably related—reduced risks require greater resources or less mobility. Of critical importance are the trade-offs that are inevitable; that is, for a given combination of risk-mobility-resources, a change in one requires change in one or both of the others.*

[2] See Selected Reading 4.3.

* This relation was demonstrated in 1974 when a reduction in resources and mobility produced a marked reduction in risk. The reduction in mobility might well have been sufficient in itself, but a simultaneous change in value system might have occurred. Certainly the savings in energy did not have a direct effect on fatalities.

The total range of goals for highway safety goes all the way from minor changes to the complete elimination of the problem. Also related to the goal is the rate of change; that is, how fast should the improvements be made. To develop first approximations of goals, the following are requirements for reasonably effective methods of reducing fatalities:[3]

Resources (cost per life saved): $100,000 to $500,000
Mobility (loss per life saved):

- Reduction of 10 million vehicle-miles of travel.
- Removal of 2000 drivers.
- Removal of 2000 vehicles.

If the growth in travel produces an annual increase in fatalities of approximately 1000, the resources required to prevent a growth in fatalities would be $100 million to $500 million per year; that is, for each 1000 reduction in annual fatalities, and for relatively effective solutions, $100 million to $500 million would be required if no reductions in mobility are made. Furthermore, to reduce the annual level by 10,000, approximately $1 billion to $5 billion would be required. These same values suggest that a reduction of 50,000 fatalities would require an expenditure of $50 billion to $250 billion. It is probable, however, that the unit cost would increase as the total annual number of fatalities gets smaller, particularly below 25, 000.

An alternative to using resources to change the system is to reduce mobility. Thus to prevent the annual growth of 1000 fatalities, 10 billion vehicle-miles of travel (or 2 million drivers, or 2 million vehicles) could be eliminated.

The time requirements are closely allied to the method that is used. Thus reducing travel could be accomplished quite rapidly, while modifying vehicles or highways could require 5 to 20 years or more.

The preceding estimates are based on the assumption that fatalities occur as a statistically random event; that is, that the removal of any driver (or vehicle, or vehicle-mile of travel) has as much chance of reducing fatalities as the removal of any other. This gross assumption can be used with care for preliminary estimates of overall goals. It does not appear, on the surface, to be an accurate assumption. For example, if 10 million alcoholics could be prevented from driving, a reduction in 15,000 fatalities could be anticipated. This would amount to reducing one fatality per 600 drivers removed. From a practical view, however, the political feasibility of preventing the alcoholics from driving is questionable, at best. For setting goals, disaggregated data can provide a good approximation.

[3] See Selected Reading 4.3.

A major difficulty in setting policy goals for highway safety is the separation of the goal-setting from the selection of the methods to be used. Oversimplifications can lead to specifying methods rather than goals, which can result in a relatively ineffective program.

CONSTRAINTS

Closely related to the goals are the constraints on the methods that are used. Constraints are of four types: resources, mobility, time, and indirect effects.

Aside from financial resources, manpower, energy, natural resources, and space also need to be considered. If limits are not set on the increase in use of resources, the technical analyses should produce estimates of whatever resources are required for the alternative techniques.

Highway accidents result from the need and desire for mobility of persons and freight. Accordingly, reducing mobility has an adverse effect on many individuals. Over longer periods of time, improved land use can lead to less need or desire for mobility. Over the short term, however, reducing mobility calls for either alternative transport systems or the loss (or reduction) of some activities, services, materials, equipment, or facilities that are provided under the existing mobility. On the other hand, benefits in addition to those from improved highway safety may accrue from a reduction in mobility.

Time is primarily dictated by political requirements. Everything being equal, the faster the improvements are obtained the better. Some methods produce rapid, but small gains. For others, the gains are slow in coming but very large when fully achieved. Usually, the more rapid the gain, the higher the cost per unit of benefit. For most political purposes, substantial gains are needed within two years, or significant progress in 2-year increments.

Indirect effects relate to nonfunctional impacts. For highway safety, these are the output of the highway system that influence the social, economic, and environmental characteristics in the environs of the highway and vehicle. Accordingly, new or greater rights of way, noise, air pollution, and neighborhood disruption have adverse effects on functions other than highway transport.

STATING SPECIFIC GOALS

By stating the goals and constraints for highway safety, the political system establishes that something *should* be done, thereby opening the door to discussions of *how* it should be done.

Sometimes the feasibility and desirability of setting goals are greatly influenced by how the gain is to be sought. Where this is known, it should be stated as a constraint. Limiting the alternatives, however, may also limit the effectiveness of the solution. But where political feasibility dictates such limits, they should be recognized as early as possible.

In light of the preceding paragraphs, goal setting for highway safety requires a political judgment on:

- Greatest feasible reduction in fatalities.
- Financial resources that can be made available.
- Possibility of reducing mobility.
- Shortest possible time for achieving a desired level of improvement.
- Any constraints that should be placed upon resource use and/or indirect effects.

Prior to obtaining technical analyses of feasibility and alternatives, the policy goal can be optimistic, and not overly restrictive.

Considering first the level of resources, there are estimates that an increase in expenditures of approximately $30 billion per year could be anticipated by the middle 1970s over the level expended for highway safety in 1966.[4] If these funds were being expended for highly effective solutions, then a savings of 60,000 to 300,000 lives per year could be achieved when the solutions were fully in effect. (Some expenditures require several years to produce full results.)

The number of highway fatalities increased from 52,924 in 1966 to 57, 000 in 1972. Progress was made, however, in the rate of fatalities, which decreased from 5.7 fatalities per 100 million vehiclemiles in 1966 to 4.5 in 1972. Thus for the amount of travel of 1972 and a fatality rate of 5.7 (1966 rate) the number of fatalities would have been 71,250. This suggests a "savings" of 14,250 lives.

An analysis of the effectiveness of the federal program[5] suggests that the actions which were taken during the 1966 to 1972 period, cost more per life saved than other methods would have cost. The level of $30 billion per year (in excess of that expended in 1966) appears to be sufficient to reduce annual highway fatalities to a level of 1000 to 10,000. What is needed is more effective actions.

How rapidly the improvement could be made is more difficult to calculate. Conservatively, 20 years would be needed. Optimistically, 10 years could be the goal, with maximum improvements possible within 5 years. To maintain popular support, a zero *increase* in annual fatalities should be

[4] See Selected Reading 4.3.

[5] See Selected Reading 4.2.

achieved within 2 years, even if this required a loss of mobility.* Furthermore, a reduction to 35,000 within 5 years could be a midprogram goal, in light of the increase in resources for highway safety since 1966.

In the 1966 to 1972 period, over 90 per cent of the funding was coming from the private sector with less than 2 per cent coming from the federal government (excluding costs of the highway construction program). Consideration of larger investments from the federal government should not be excluded, even though such investments may seem politically unattractive and perhaps unnecessary.

For purposes of this discussion then the goal statement for highway safety is as follows:

- The annual number of highway fatalities (*a*) will not increase (within 2 years), (*b*) will not be greater than 35,000 (within 5 years), and (*c*) will not be greater than 5000 (within 10 years).
- The maximum annual expenditure of funds (over the 1966 level) will be $30 billion, with the source of funds being the private sector, unless more effective solutions are possible with state or federal funding.
- Losses from a reduction in mobility will be offset by new transport capability, or by other means.
- Solutions that reduce use of other resources (manpower, energy, materials, equipment, space) should be sought.
- Indirect effects on the environs will be equal to or better than those that existed in 1973.

This policy statement, explicitly or implicitly, establishes the following political judgments:

- *Total number of fatalities* is the primary social value with which the political system is concerned.
- Action *should be* taken to reduce highway fatalities.
- Although there is a relatively low potential for improving life expectancy through a reduction in highway fatalities (as compared to programs to reduce fatalities from heart disease and cancer, for example), the benefits can be substantial.

* The improvement in 1974 was achieved at no increase in cost for automobile travel, but it required a reduction in mobility. The change also illustrates the principle that nearly 100 per cent of the vehicles and/or drivers need to be changed or influenced in order to have a rapid change in highway fatalities.

- Achieving the 10-year goal would reduce the consequences of highway accidents to a level that is comparable to or less than the number killed by accidental falls, by drowning, or from fires, and would be one-third of the number killed in work accidents.

Policy Analysis

From Figure 2.5, the steps in policy analysis are identified as:

- Development of criteria measures.
- Identification of societal system sensitive to goal.
- Identification of alternative means for meeting goals.
- Evaluation of alternatives relative to performance criteria.
- Evaluation of feasibility of alternatives.

DEVELOPMENT OF CRITERIA MEASURES

System Output. Highway fatalities are one of the "prices" paid for the movement of persons and goods; that is, they are one measure of the quality of service of highway transportation. A reduction in highway fatalities requires changes in one or more of the characteristics of the output of the highway system. The quality of service of highway transportation is classified in Table 4.4.

The needs and desires of individuals and social systems are the source of the quantity and quality requirements, tempered by the level of resources which are available or applied. The larger the *quantity* of transportation, the better the transport system, but not necessarily the social system. The quality of the social system is based on many factors, only some of which are advanced through transportation. The quantity of transportation in a social system is a function of (1) the quality of service that is provided, and (2) the benefits or adverse effects that are derived. The benefits and adverse effects from a given quality of service are the determinants as to the ultimate quantity of transportation that is accomplished.

Analyses of alternative means for reducing highway fatalities should not oversimplify the fact that highway fatalities occur because of the individual needs and desires for transportation. Since (1) the quality of service influences the quantity, and (2) the quantity influences other social goals, a clear idea is needed as to how much influence a given highway safety measure has on the output of the highway system. The following section discusses the influences of the quantity and quality of highway transportation on other societal systems.

Table 4.4 Quality of Service of Highway Transportation

1. *Quantity*
   - a. Ton-miles of freight movement
   - b. Passenger-miles of personal movement
2. *Quality*
   - a. Cost
     - (1) Total (all resources)
     - (2) Cost per ton-mile
     - (3) Cost per passenger-mile
   - b. Time
     - (1) Total (origin to destination)
     - (2) Average speed
     - (3) Reliability (time loss)
   - c. Safety
     - (1) Total number of accidents
     - (2) Total severity of accidents
     - (3) Accidents per vehicle-mile of travel
     - (4) Severity per vehicle-mile
   - d. User Satisfaction
     - (1) Facilities
     - (2) Vehicles
     - (3) System operation
     - (4) System administration
   - e. Indirect Effects
     - (1) Physical (air pollution, noise, etc.)
     - (2) Psychological (esthetics, neighborhood or areal disruption, etc.)

For safety, the major highway system criteria deal with rates of loss per vehicle-mile of travel, and the most frequently used rate is the number of fatalities per 100-million vehicle-miles. In 1972, for example, the rate of fatalities was 4.5 per 100-million vehicle-miles. To reduce the annual number of fatalities to 5000 would require a rate of 0.4 (assuming zero growth in travel). If the quantity of travel was cut in half, the rate would need to be reduced to 0.8—still a substantial reduction.

For perspective, without otherwise changing the system (i.e., a fatality rate of 4.5), to reduce fatalities by reducing travel would require a 90 per cent reduction to 110 billion vehicle-miles, which was the level of highway travel in the early 1920s.

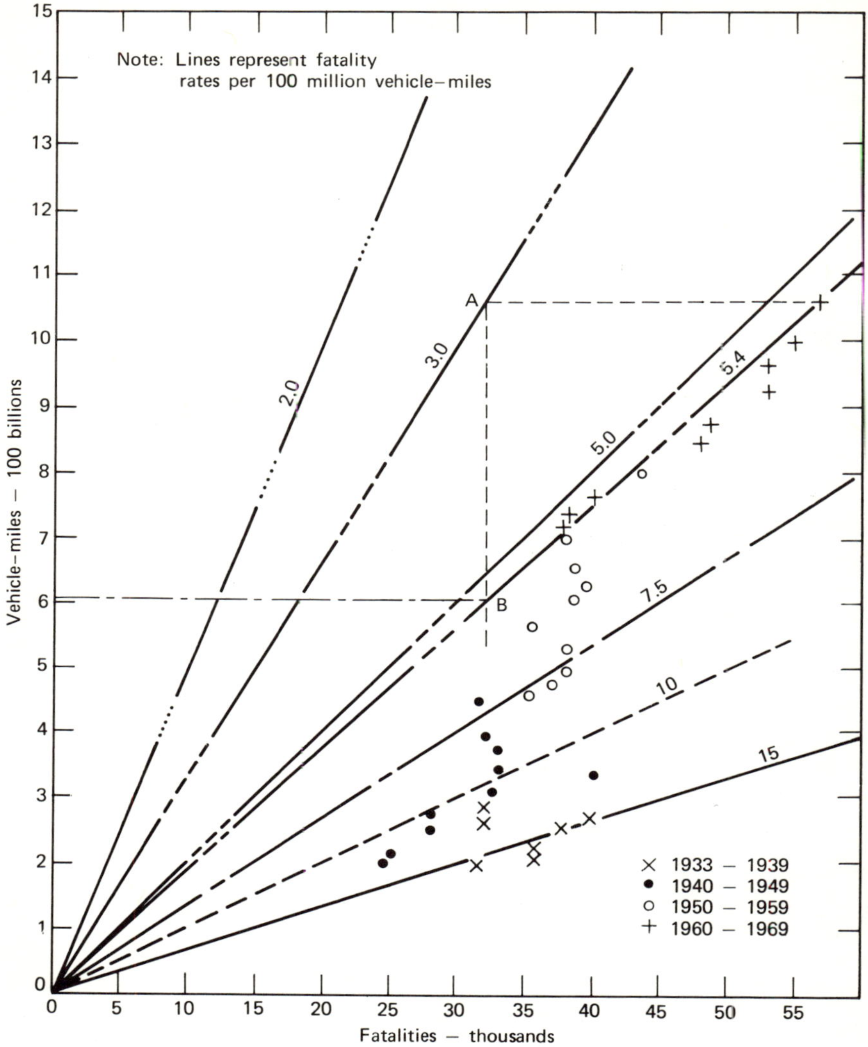

Figure 4.2. Risk and fatalities. *Source:* Selected Reading 4.3, p. 161.

The rate, therefore, is an important criterion from the viewpoint of system modification (Figure 4.2). Used in conjunction with positive and negative changes in the amount of travel, various alternatives can be developed.

Economic Effects. The economic consequences are an important criterion. Of primary interest is whether there is likely to be a net economic loss or gain in operating the highway system because of the implementation of an alternative. The analytic technique that is usually used is the benefit/cost ratio.

To develop an estimate of the economic costs and benefits to highway transportation, only those factors that have a direct monetary effect should be included; that is, costs will include those resources that have a monetary value, and the benefits will be *reductions* in the costs of operating the existing system. Nonmonetary costs and benefits should be identified, but not included if they do not influence fiscal accounts. This is not because the costs and benefits may not be "real," but because the purpose of the economic analysis is to establish whether there is a monetary gain or loss from the investment. A subsequent analysis clarifies the overall social effects.

All resources that are used, however, should be identified, and the quantities listed for subsequent analyses. For example, land, energy, and manpower are important resources in their own domain, aside from their monetary costs. As shortages in land, energy, manpower, and other resources develop, how *all* resources are used becomes of significant interest.

The economic criteria that are needed then include:

1. Total monetary cost (and possible sources).
2. Total monetary benefits.
3. Time relations for expenditures and for benefits.
4. Benefit/cost ratio over time.

For the highway safety goals it was not specified that the benefits should exceed the costs. This is because the goal was a social one, rather than economic. However, if the benefit/cost ratio is equal to or greater than one, resources in addition to the $30 billion would become available; that is, the policy goal did not require that the $30 billion per year be recovered from transportation savings.

As noted in a subsequent section, some measures produce other benefits than simply a safer system. For example, a new highway, normally with a benefit/cost ratio of at least 3, produces a savings in both lives and costs over the life of the road. The latter results because of monetary benefits from reduced operating costs. The net effect is that there is a zero cost per life saved.

Cost Effectiveness. The objective is to reduce fatalities with specific limitations on resources, time, and indirect effects. If the goal can be met with less expenditures, then this would be a preferred solution. A technique for analyzing the relative effectiveness of an expenditure is termed cost effectiveness (discussed in Chapter 3).

The two major considerations are (1) the total benefits and (2) the cost per unit of benefit. Total effectiveness, for highway safety, means the total reduction in loss of life, while the rate is expressed as the cost per life saved. For a given alternative then the total number of lives saved is an important criterion. The rate is critical to develop the combination of several actions that save the most lives for a given investment. For example, if one method costs $500,000 per life saved, and another costs $5 million per life saved, then it is apparent that the more expensive alternative would be a waste of money insofar as saving one life is concerned.

Neither the total number nor the rate is sufficient by itself. Although the total may answer the question of whether the savings is or is not large enough, it does not provide insight on the resources required. The rate is not enough, because it does not provide data on the total number that could be saved.

Cost effectiveness usually deals with monetary costs and time, and indirect effects are not necessarily factors. For highway safety, these simplifications are not adequate. *All* resources should be considered individually to see if there is a significant use of a resource per life saved. Time and indirect effects are also critical because of the goal statement. Indirect effects can be considered in the same way as monetary or resource costs are considered; for example, positive or negative increases in air pollution per life saved.

As noted earlier, in the United States in the early 1970s, relatively effective corrective measures cost in the range of $100,000 to $500,000 per life saved. Accordingly, efforts should be made to identify techniques with this level of effectiveness. The 10-year goal of 5000 fatalities per year, however, may well require costs in excess of one million dollars per life saved.

IDENTIFICATION OF RELATED SOCIETAL SYSTEMS

Three types of societal systems or functions have an interest in the actions taken for highway safety:

- Those with health and safety goals.
- Those that provide health and safety services.
- Those that use the highway.

Systems with Health and Safety Goals. In the preceding pages, comparisons have been made between causes of death. In addition to heart disease and cancer, there are many other causes of death at which corrective programs could be directed. To persons or organizations with a primary goal of increasing life expectancy, programs to reduce highway fatalities are of interest.

Certainly some resistance, or at least competition, can be expected, particularly if the expenditures are not too effective. If the same expenditures in other programs could produce a greater savings of life, then the political wisdom of the highway safety expenditures is likely to be challenged.

The validity of these challenges depends on both the total number of individuals that other programs could influence and the relative effectiveness of expenditures in the alternative programs. Therefore, if the potential for saving lives is not as great, and/or if the cost to save a life is significantly greater, then there is little basis for a claim of misuse of resources on highway safety.

There is also the aforementioned cultural characteristic that views some deaths as being worse than others. As discussed earlier, "useless" or "unnecessary" deaths are particularly deplored, as are events that produce multifatalities. This attitude may or may not be rational. Where it exists, it can have a dominant effect. It is likely that the idea of saving 700,000 lives per year through minimizing heart disease is not a popular goal both because such deaths are not being viewed as unnecessary or as easy to reduce, as highway fatalities are viewed.

Conferences and joint analyses with persons in the medical profession are helpful in technical studies to determine how highway safety fits into the total program for increasing life expectancy. The criteria used by these health and safety societal systems are almost exclusively health and safety related. Their usual view of transport requirements is that they are irrelevant to the primary goals of life expectancy. Accordingly, their criteria for solutions are likely to be based upon rates of fatalities by population. In the early 1970s, this rate for highway safety was in the range of 28 per 100,000 population.

For this criterion, reduction of travel and passenger protection is equally or more desirable than techniques to improve safety through developing a better quality of transportation.

Health and Safety Service Systems. Emergency services, including the related activities of police, fire protection, communication, and transportation, have a direct interest in the highway safety problem. These services are not limited to responses to highway accidents. Rather, they are concerned with all unexpected health and safety crises of individuals and groups.

The primary goals of these services are stated in terms of life expectancy; that is, their objective is to save lives and to reduce the severity of injuries. Thus they are closely allied to the other societal systems whose attentions are directed to health-related problems.

The criteria for emergency services are stated in number of lives saved, and probable reduction in hospitalization, permanent or partial disability, and/or days spent in recovery. Rates of losses are important variables; that is, the severity per occurrence is an important measure of the adequacy of the service.

Prevention is a major goal of these services. Through education, public relations, and consumer protection media, efforts are made to remove "causes" of accidents. Total losses and losses per 100,000 population are also important measures that reflect successes or failures in reducing the number of emergencies each year.

Their interest in highway safety then is equally directed to the reduction of the number of accidents and to the severity of personal injury. In addition to accident prevention, they are interested in the protection of individuals who are operating or riding in an automotive vehicle.

Finally, a rapid response to calls for assistance is an important consideration. Prompt treatment is a major factor in reducing the consequences of an injury. Therefore, time response per emergency is an important criterion.

Emergency services are concerned about all types of accidents. From an objective base, their goal is not particularly affected by the cause of the emergency. The potential for their improvement is great in highway safety, of course, and equal to that from all other types of accidents. But many emergencies arise from illnesses, from criminal assaults, and from other origins in the home or office. Therefore, it is uncertain that highway accidents offer a particularly fruitful sphere of activity in which to seek improvement of such services.

The cost effectiveness of measures related to emergency services are relevant. One attribute could be the simultaneous improvement of life expectancy in other areas through development of systems that provide assistance not only to those involved in highway accidents but also to those involved in other emergencies as well. While certain techniques and actions are uniquely suited for responding to highway accidents, it is quite possible that for minor additional expenditures much more cost effective approaches to improving life expectancy could be developed.

Accordingly, societal systems that provide emergency services are interested in the following criteria: (1) casualties per 100,000 population, (2) severity per accident, and (3) total losses. Their interest is more marginal in the number of accidents per vehicle-mile of travel, but as an index of the effectiveness of prevention, accident rates based on travel are relevant to their goals.

Systems That Use Highway Transportation. Individual use of highways is the usual method for analyzing highway requirements because of the simple, direct measures that this approach makes possible. But transportation is the means to the end for many functions, and the quality of highway service influences a broad group—much more widespread than simply the vehicle operator and/or the passengers on a given trip. The fact is—all societal systems depend to some extent on highway transportation, both of persons and of freight, to achieve their goals.

For the industrial, cultural, recreational, and public service systems that require highway services, but do not provide a part of the transport function, the costs of highway accidents are of primary interest. Industry and public sector agencies that operate truck fleets are concerned about all losses that occur to their employees and to their equipment and materials. All groups are affected by loss of employees who are casualties of highway accidents.

As users of highway transportation, all societal systems are influenced by the benefits and cost increases that result from safety measures. Reductions in travel also result in significant social and economic losses. These losses can occur both in the private and the public sectors. Therefore, systems that use highway transportation have a different interest in the actions taken to improve highway safety than do the societal systems concerned with health and life expectancy. As users, they are interested in the loss per vehicle-mile of travel, for example, because of the direct effect on their operations. For those systems that do not operate vehicle fleets, increased costs of freight have an influence on the effectiveness with which they can achieve their own goals. This is offset by the reduction in lost time (or lost employees) from personal injuries.

Administrators of highway user systems are more realistic about highway accidents than are those concerned with life expectancy goals. This is because the risk as it is perceived is more than offset by the benefits from the transport. The amount of protection that these users wish to have provided is controlled by the resources that they have available to devote to that purpose. Since users have specific objectives of their own, there is a limit to what they wish to spend to reduce highway risks. For example, a school system might not choose to buy new school buses, not because they are not concerned about the safety of their children, but because they may also need safer or better school buildings, better equipment, or higher salaries for teachers. This is not necessarily an inhumane decision.

In summary, the criteria of interest to highway system users are measures of the rates of loss or risks of use of the highways. Total losses on the highway, being a function of the amount of use, should be of less interest than the risk per unit of use; that is, safety per mile of travel is of more consequence than safety through less travel. Further costs to achieve greater highway safety may or may not be accepted as an effective expenditure of resources insofar as their mission is concerned.

Policy analyses need to develop output measures of highway transportation to provide a more complete understanding of the impact of the highway change on other societal systems.

IDENTIFICATION AND EVALUATION OF ALTERNATIVE MEANS

Table 3.2 lists the elements of societal systems that can be changed. For highway safety, these elements or components can be identified as:

- Highway facilities
- Vehicle
- Driver (or control system)
- Emergency services

To reduce highway fatalities then a change or changes *must* be made in the characteristics or behavior of one or more of these four variables.

The goal statement that is to be evaluated includes the requirements that fatalities be reduced under the following time schedule: (*a*) zero increase during and following the third year; (*b*) no more than 35,000 fatalities per year after the fifth year; and (*c*) no more than 5,000 fatalities per year after the tenth year. A total of $30 billion per year is assumed to be available, including funding for programs initiated by the 1966 acts.

The Existing Situation. As noted previously, highway fatalities are one of the outputs—one of the prices that is paid—for highway transportation. By the mid-1960s, there were few responsible officials or safety authorities who believed that all fatalities could be eliminated. No one but crusaders, wishful thinkers, and/or uninformed individuals believed that a relatively simple, rapid solution was available. Because of the massiveness of the highway-vehicle-driver system, and because of the dependency (and some said overdependency) on highways for transportation, rapid improvement was obviously difficult or impractical.

Tables 4.5 and 4.6 illustrate the size and complexity of the highway system. To improve one or more of the four basic elements—highway, vehicle, driver (control system), or emergency services—a massive, cumbersome system must be influenced. For modest improvements, changing any one of the four could produce results. But for any major reduction in

Table 4.5 Trends of Highway Safety Variables*

		Annual Increases			
	1966	1962–1966	Percentage of 1966[a]	1966–1970	Percentage of 1966[a]
Highway system quantity criteria					
Travel (in billion vehicle-miles)	930	43.5	*4.6*	51.0	*5.7*
Number of vehicles (in millions)	95.9	4.0	*4.2*	3.9	*4.2*
Number of drivers (in millions)	101	3.6	*3.6*	3.6	*3.6*
Safety criteria					
Accidents (in thousands)	13,600	650	*4.8*	600	*4.4*
Injuries—disabling (in thousands)	1,200	100	*6.8*	25	*2.9*
Fatalities	53,041	3060	*5.8*	490	*0.4*
Property damage (in millions—1966 dollars)	$ 3,300	$ 200	*6.1*	$175	*5.3*
Total economic—property damage, medical fees for injuries, insurance administration (in millions 1966 dollars)	$10,000	$ 0.75	*7.5*	$ 0.37	*3.7*

Source. Selected Reading 4.2. Based on arithmetic averages from data contained in *Accident Facts*, 1962–1970 Editions, National Safety Council, Chicago, Illinois.
[a] Percentages are in italics.

Table 4.6 Approximate Values for Estimating Effectiveness of Highway Safety Programs (Late 1960s)

Drivers	
100 million	Total drivers (50 per cent of the population)
2 million	New drivers per year
6.5 million	Alcoholics with driver licenses
11 million	Drivers under age 20—and driving in 17 per cent of the fatal accidents
9 million	Drivers over age 64—and driving in 10 per cent of the fatal accidents
32 million	Drivers under age 30—and driving in 45 per cent of the fatal accidents
Vehicles	
100 million	Total licensed vehicles
10 million	New vehicles each year
10 years	Average life of vehicle
10 percent	New vehicles with safety defect
20 percent	All vehicles with safety defect
55 m.p.h.	Average vehicle speed—rural
15 percent	Vehicles less than 2 years old
25 percent	Vehicles 2 to 4 years old
60 percent	Vehicles more than 4 years old
Highways	
3.7 million	Total miles
41,000	Miles on completed interstate system
885,000	Miles with federal aid
2.8 million	Miles without federal aid
Services and administration	
33,000	Number of government agencies with responsibility over some highway mileage
30,000	Total number of highway patrolman
$25,000	Cost per patrolman
$20 billion	Annual expenditure for highways (all levels of government)
$5 billion	Annual expenditure for construction of interstate system
$160 billion	Annual expenditure for highway transportation (20 per cent of the GNP)
Traffic and safety records	
1,700	Average annual increase in number of fatalities
37 billion	Average annual increase in vehicle miles
99.93	Percent of drivers not in a fatal accident
99.93	Percent of vehicles not in a fatal accident
20 million	Approximate number of vehicle miles of travel per fatality

Source. From Selected Reading 4.3, p. 14.

fatalities, it is probable that suboptimization would result if only one of the four were to be changed. Here suboptimization or "gilding the lily" means improving one element too much. More effective results would be achieved if each element were to be improved in a manner and to the extent that the overall system performs in an improved manner.

Fatalities had been and were occurring in a highly random manner throughout the highway system. The most reliable relationship was between the vehicle-miles of travel (or exposure) and accidents. There was convincing evidence, also, that fatalities were related to certain parts of the driver population; that is, alcoholics, other drivers under the influence of alcohol, and the youngest and the oldest age group. Data on groups of drivers were limited, however, particularly on the amount of vehicle-miles of driving per group.

Table 4.7 shows the random character of highway fatalities. Even though fatalities and vehicle-miles appear related for aggregated national data, there are major discrepancies for disaggregations. For example, there were three times as many nonpedestrian fatalities in rural areas as there were in urban areas, even though the total vehicle-miles of travel was approximately the same; that is, the fatality rate was three times as high in rural areas. On the other hand, there were nearly twice as many pedestrian fatalities in urban areas, making this fatality rate nearly double that in rural areas—again for equivalent travel.

The expenditures per year for highway safety for the 1966 to 1972 period were difficult to obtain in 1973. Ranges of these expenditures were available, however. As noted earlier, if all the highway safety standards contemplated and promulgated by the Executive Branch on the basis of the Federal Highway Safety Acts are fully implemented, then approximately $30 billion per year (over the 1966 level) may well be expended.

The benefits to be derived from this federal program were stated[6] as a reduction in the fatality rate to a level of 3.6 per 100 million vehicle-miles by 1980, with the prediction that even with this reduction in rate, there would be "over 50,000 fatalities" per year in 1980. The reduction from 5.7 (the rate in 1966) to 3.6 would mean that for 1980 the fatalities would be approximately 60,000 rather than the 100,000 for the 1966 rate (assuming that the average growth of 5 per cent in vehicle-miles of travel was to continue). Assuming zero rate of traffic growth, the projections for 1980 would be 45,000 and 71,500 fatalities, respectively, for the 3.6 and 5.7 fatality rates.

For the savings of 40,000 (or 25,000) lives in 1980, and an annual expenditure of $30 billion, the cost per life saved would be $750,000 (or $1.2

[6] See Selected Reading 4.10, Safety Report, 1971.

Table 4.7 Typical Statistics for Estimating Limited Effects Benefits of Highway Safety Measures

	Percentage of Fatalities	Number in 1968	Type of Accident	Percentage of Fatalities	Number in 1968
Pedestrians	18.0	9,800	Railroad crossing	3	1,630
			Number of crossings		225,000
Motorcyclists	3.5	1,900			
Trucks	9.3	5,150	Day	47	25,900
			Night	53	29,300
Commercial buses	0.2	130			
School buses	0.05	30	Urban	32	17,500
			Urban—over 1 million	4	2,350
Passenger cars	65.0	36,400	Rural	68	37,700
Emergency vehicles	0.3	180	Limited-access turnpike	1	608
Stolen car	0.2	100	Interstate system	6	3,500
			County roads	16	8,800
			Head-on collision	17	9,300
			Rear-end/angle collision	16	8,800
			Single vehicle	31	17,000
			Wet pavement	16	8,800
			Snow and ice	3	1,600

Source. Selected Reading 4.3, p. 19.

million). This cost per life saved is higher than some estimates have suggested should be possible, but this cost rate is not unreasonably high.

General Alternatives. The programs and goals of the 1966 to 1972 period do not meet the policy goals as stated for our example. Therefore, a major change would be required to achieve the goals of the example. Returning to the three goal statements, the following observations on general alternatives can be made.

1. *To assure no further increases in fatalities within 2 years.* Assuming that travel increases at the rate of 5 per cent per year, and assuming no changes in the system, the annual increase in vehicle-miles of travel is approximately 50 billion vehicle-miles, and approximately 2000 fatalities. This calls for a reduction in fatality rate from a level of 5.00 to 4.75. To produce this type of improvement within 2 years, including time required for legislation, the following observations can be made:

a. *Highways.* Highways are built and maintained by the states. With typical lead times and construction periods, it is impractical (or impossible) to achieve any significant shift over what is currently underway. Accordingly, there is a very low probability of achieving the goal by changing the highways within the time frame allowed.

b. *Vehicle.* There are 100 million vehicles on the road and 10 million new vehicles each year. Assuming at least 6 months for a production change, approximately 20 million vehicles could have an influence within 2 years. This constitutes approximately 20 per cent of the vehicle fleet. If this 20 per cent were to have its proportional share of the potential for fatal accidents, then it would need to produce the savings of 2000 lives. This would be equivalent to a total savings of 10,000 lives (if the entire fleet were equipped) and a reduction of the fatality rate to the level of 4.0 (for 1972 traffic).

As to the equipping of the existing 100 million vehicles (retrofitting), given sufficient priority, and relatively unsophisticated changes, it is probable that retrofitting could be accomplished in time to achieve results within the desired time frame.

Therefore, modification of the vehicle or the removal of vehicles from the road cannot be exluded on the grounds of being impractical or impossible within the given time frame.

c. *Driver (control system).* There are over 100 million drivers, and approximately four million new drivers each year under 1973 conditions. The status of driver education and the potential for change preclude any reasonable chance that capability for improving driving technique can be helpful within 2 years.

Control of the driver through police and other administrative action is another matter. Three of the more obvious possibilities include:

- Removal of drivers' licenses of a group (alcoholic, "high-risk," etc.)
- Restrictions on or rationing of vehicle operation
- Policing or other changes of traffic operations.

The feasibility or effectiveness of these measures is questionable, but control of the driver is certainly a feasible alternative for the short-term goal.

d. *Emergency services.* These services are provided for the most part by the private sector. They are fragmented, dispersed, and subject to many demands other than those for highway safety. For the short term, considering the legislation and fiscal requirements, it is impractical to impossible to expect significant contributions to the desired reduction in fatalities.

2. *To reduce the annual level of fatalities to 35,000 within 5 years.* The 5-year goal, while ambitious, could be achieved in a number of ways. As with the 2-year goal, however, the lead times are a serious obstacle as would be any requirements for modifying *all* highways, *all* vehicles, *all* drivers, and *all* emergency services.

Each of the elements could be substantially changed within the 5-year period, even with 2 years devoted to preparations for implementation.

Certain general observations are possible, relative to achieving the 5-year goal:

- Changing *only* the highway is *not* feasible.
- Changing *only* the vehicle is *not* feasible.
- Changing *only* emergency services is *not* feasible.
- Changing *only* the driver is feasible, but *only if* substantial reductions in travel are a part of the solution.

The most effective solution for reaching the 5-year objective is through changes in more than one of the elements, that is, overall system inprovement. The goal will be difficult to achieve and may be impractical within 5 years without a reduction in mobility.

3. *To reduce the annual level of fatalities to 5000 within 10 years.* A reduction of this magnitude is nearly equivalent to the "elimination" of highway fatalities. The 5000 deaths per year would still be substantial—comparable to the casualties from drowning (7500), fires (6700), poisoning (3000), and firearms (2300). This amount of reduction from a level of 50, 000, however, and the much greater exposure of individuals to highway accidents, than to the other listed activities, would require an effort that would be similar to one for eliminating the problem.

The following paragraphs examine the extent to which the elements can be changed to produce the desired reduction.

Highways. The fatality rate for the highest quality of highway—limited access facilities—was 2.8 in 1970. For 1980 travel, if this fatality rate existed for *all* highways, there would be approximately 50,000 fatalities. Thus even if such a change were socially, economically, politically, or technically feasible, the desired reduction in fatalities would not be achieved by *only* changing the highway.

Vehicles. It is conceivable that they could be so constructed as to practically eliminate fatalities. Whether this could be accomplished within 10 years is highly questionable, however, because at least 5 years would be required for the cycle of research, development, testing, evaluation, production changes, and public acceptance. Accordingly, it is probable that half of the vehicles (the older style) would need to be eliminated from the highway to achieve the goal in 10 years.

Other changes in the vehicle (tires, mechanical condition, maintenance, etc.) that would reduce accidents would not completely overcome the critical role of the driver. Accordingly, changes in the vehicle would also need to include changes in the control system and/or the highway (guideway).

Even if the vehicle were made "fatality-proof" for its occupant, there would still be 10,000 pedestrian fatalities to consider. Modification of vehicle exteriors to prevent pedestrian fatalities appears to be practically impossible, even though some improvements appear to be practical.

Drivers. There is no reason to hope that their performance with the existing quality of vehicles and highways can be improved sufficiently to reduce fatalities to 5,000 per year. Driving technique can be improved. Some if not all of the highest risk drivers can be removed from the highway, but to provide the type of improvement that would be necessary would take an unprecedented level of human behavior.

Emergency services. Improvements here alone could not provide the necessary reductions in fatalities. The character and severity of many fatal accidents means that more direct efforts at prevention or occupant protection would also be necessary and might in themselves be more effective.

In other words, there is reason to doubt that the 10-year goal is feasible. There is even greater question as to whether any technical solution would be socially and politically feasible. Certainly the goal is a socially desirable one, and the technology no doubt exists to provide alternative solutions.

Summary. To reduce fatalities to 5000 per year one or more of the following is essential: (1) the "control system" needs to be improved to where the error is only one fatal judgment per 20 billion that are not; (2) vehicle occupants need almost perfect protection; and/or (3) mobility needs to be substantially decreased.

Table 4.8 Steps for Producing Estimate of Relative Effectiveness of Highway Safety Measures

1. Identify specific safety measure
2. Conduct benefit analysis
   a. Determine number of units involved—that is, vehicles, drivers, miles of highway, number of grade crossings, etc.
   b. Determine "life" of correction—that is, number of years it will be effective without replacement
   c. Determine number of fatalities (or other measure) that the measure can influence
   d. Estimate the range of effect of the proposed safety measure on the annual number or per cent of fatalities—that is, a low and high estimate of benefits per year is developed
   e. Estimate the effect of the measure for the life of the measure
3. Conduct cost analysis
   a. Estimate capital costs
   b. Estimate maintenance and operation costs per year for the life of the measure
   c. Estimate total cost, including interest on the investment if applicable
4. Develop rate of implementation
   a. Estimate time for legal and administration actions
   b. Estimate time to manufacture, construct, distribute, etc.
   c. Adjust cost and benefit estimates for the time requirements to produce results
5. Develop measures of effectiveness
   a. Total benefits within time requirements
   b. Total cost within time requirements
   c. Unit measures of effectiveness—that is, cost per life saved, cost per accident saved, cost per measure of accident severity, etc.
6. Comparisons with other measures
   a. Combinations of measures that will produce desired results within the time requirements
   b. Costs for various combinations
   c. Special problems of implementation—that is, legal, administrative, political

Source. Selected Reading 4.3, p. 153. Based in part on unpublished report by Harry Weingarten and David Solomon, U.S. Bureau of Public Roads, August 1967.

An objective appraisal of alternative solutions to achieving the policy goals requires procedures that are unique to highway safety. Similarly, each system that undergoes a policy analysis has a set of special considerations that requires analysts who are fully familiar with the components and operation of the system, and how it can be modified to achieve the desired result. System modifications call first for a sound technical analysis, followed by a policy decision. The following section examines the methodology and then considers the evaluation.

Methods of Analysis of Alternatives. Estimating relative effectiveness of various measures uses a conventional cost effectiveness analysis. Table 4.8 outlines one such approach. The most troublesome feature of the evaluation is the prediction of the benefits to be derived, particularly the reduction in fatalities. This problem involves two types of judgments that call for professional skill. One is the probability that a corrective measure, if fully implemented, would reduce fatalities a given amount. The second is the probability that a given measure can be fully or partially implemented. The first is primarily a technological problem, while the second requires both technological and social-political judgment. The latter is excluded from the technical evaluation, and discussed under the following section on feasibility.

Judgments on reduction in fatalities require highly professional, unbiased estimates. To minimize the effect of extreme views, the consensus of a group of experts should be obtained. Special techniques for eliciting such judgments have been developed. By a skillful selection of the questions for which judgments are requested, the variation in the estimates of fatalities can be substantially reduced. Thus information on the probability of reducing tire blowouts by specific measures is relatively easy to obtain from an expert on tires. But the conversion from the probable total number of blowouts to fatalities needs to be made by a different expert—one who is more familiar with traffic operations. The final result is less variable and more reliable than estimates of reductions in fatalities by either an expert on tires, or one on traffic, or a mixed group.

Realistic time tables and annual cost requirements are also essential parts of the effectiveness analysis. These analytic techniques are discussed in Chapter 3.

Given the costs and the benefits, the cost per life saved and the total lives saved per year can be estimated. The results from analyses of various measures are subsequently combined to obtain various combinations of actions that produce the desired objective.

Redundancy is a special problem, because the benefits to be derived from any single corrective measure depend on other actions that are taken. For example, if mobility is reduced, the benefits derived from any other corrective measure are also reduced.

Correspondingly, if accidents are prevented, the benefits from purely protective measures are reduced. Thus a reduction in the number of accidents reduces the benefits that are derived from the use of seat belts.

Other redundancies are actions that are directed to the same type of loss. For example, collapsible steering wheels and seat belts are addressed to driver protection. Either one is more effective than none, but the combined effect is less than the sum of the benefits from either one used alone. Similarly, crash protection at highway obstacles reduces the benefits from driver or passenger protection, and vice versa. Some redundancy is desired, but in estimating benefits, the effect of the duplication should be taken into account.

Protection of occupants versus prevention of accidents is as valid a debate in highway safety as it is in fire fighting. The two approaches call for totally different actions. The benefits derived from the protection of vehicle occupants can be quite different from those obtained from the prevention of accidents. Protection has a relatively nonproductive effect on transportation; that is, there is only a minor improvement in the quality of service. Furthermore, the cost effectiveness of protection is directly comparable to that for other means of improving life expectancy. The special benefits to transportation from improved safety are usually paralleled by the benefit from savings of life or reductions in injuries on any function.

For accident prevention, some corrective measures produce improved transportation, either in time savings or in operating costs. These benefits are in addition to those obtained from accident savings alone. Thus increased speed and more uniform traffic flow produced by the removal of horizontal curves and reduction of steep grades can result in substantial economic savings in addition to social gains from accident reduction. In fact, the improvement of a 3- to 5-mile section of highway is normally expected to produce benefits that are three times the cost. Since accident savings are a relatively small part of the benefits, the ultimate effect is a savings in life at zero cost. Highway improvements do require an investment as a means of reducing fatalities, but on an annual cost basis, these expenditures are recovered through reduced operating cost.

The preceding observation of annual cost for prevention of accidents is not valid for measures that are purely for the protection of human life. It is valid, however, for consideration of economic savings for protective devices to reduce property damage.

It is also valid, of course, to show the economic benefits of life preservation through highway safety measures. The results, however, are of use only for demonstrating that overall economic benefits are derived from the highway safety programs; that is, as a justification that something *should* be done. Economic benefits from preservation of life do not help in solving the problem of *how* highway safety should be improved, particularly when the goals are stated as reductions in fatalities.

Feasibility of Alternatives. The feasibility of the alternative courses of action consists of technical, economic, social, and political considerations. These analyses consist primarily of technical judgments from experts in related fields.

The technical feasibility of the proposed 5- and 10-year alternatives requires an evaluation of the following:

- Can vehicle-based control systems be developed within 5 years? If so, what are the costs for research and development, testing and evaluation, investment for production, and cost of the equipment?
- Can a "fatality-proof" vehicle be developed within 5 years? If so, what are the costs for research and development, testing and evaluation, investment for production, and selling price of the vehicle?
- Can alcoholics be identified? Can social drinking and driving be controlled? Can alcohol and driving be eliminated? Will this effectively increase life expectancy, or will other problems be multiplied?
- Will the removal of the youngest or oldest age groups reduce fatalities more than the removal of any similar number of drivers? If so, how much? Will this produce other problems that offset the gains in highway safety?

Similarly, each alternative is evaluated by experts who can make technical judgments. Some estimate of the reliability of the alternative should also be developed.

Economic feasibility deals with the economic benefits and costs of the alternatives. The objective is to determine whether there is a net economic gain if the alternative is implemented. Existing and potential sources of revenue or expenditures should be identified. The timetable for economic receipts, expenditures, and benefits should be developed, and if a subsidy is required, the amount should be stated.

Social feasibility deals primarily with individual acceptance or rejection of the alternative. It also considers the impact on related programs. If a reduction in mobility is involved, the social implications should be considered. The primary objective is to establish how well an alternative fits into existing and future social systems.

Political feasibility considers whether it is politically possible to implement an alternative. There are questions on taxation, on the reaction of individuals to legislation or executive action, and on the ability of the political system to retain public support if an alternative is implemented.

Feasibility analyses do not attempt to make political judgments, but to present objective appraisals. From them, recommendations for the best course of action can be developed. Again, however, the assumptions that underlie a recommendation should be identified.

The reports of the technical evaluation and the feasibility analyses are returned to the policy-makers for review and action. For the goals advanced for this example, a reasonable alternative is to revise the goals prior to implementation. For example, the 10-year goal could be stretched out to 15 or 20 years which would reduce the cost and minimize the possibility of removing outdated vehicles.

If the goals are changed, the policy analysis cycle should be repeated, and a new set of alternatives should be prepared.

The Decision

The final decision from the policy-makers includes the specific goals and constraints, and identifies the alternatives by which the goal is achieved. The sources of revenue should be identified, and necessary papers for legislation need to be prepared.

To become policy, legislative action is required, including tax measures (if these are necessary). The goals are most easily achieved and with the highest probability of success if the legislation is as specific as possible on the goal and the constraints. Overly rigid legislation on the methods to be followed, however, reduces the cost effectiveness of the ultimate solution.

For the preceding 10-year goal, the number of fatalities and resources available should be specified. Furthermore, the specification of an improved control system rather than occupant protection can be made. The decision on the details of the control system (e.g., its being vehicle contained, based on use of radar, and other such specific characteristics) should be delegated to the Executive Branch.

The 2- and 5-year goals should also be covered in the legislation, along with the constraints. Since reasonable alternatives exist, the method should be left as flexible as possible. Preferably, a limit could be placed on the cost effectiveness of the technique. Some local areas, for example, might prefer one form of travel restriction to another.

To reduce problems of redundancy, the actions which are implemented nationally should be clearly identified.

Guidelines for setting standards for highway, vehicle, driver, and emergency services should also be carefully stated. Once again, a cost effectiveness analysis minimizes the waste of resources on a change that has marginal or no effect on fatalities.

EVALUATION OF POLICY DECISIONS

In Figure 3.1, the feedback to the social system is shown. For national policy development, this feedback should include an evaluation of how well a given policy has worked. For the highway safety problem, an evaluation of the effect of the 1966 Federal Highway Safety Acts can be used as an example.[7]

The 1966 Acts did not provide specific goals, but called for "assistance and cooperation" to "increase highway safety"; and furthermore "to reduce traffic accidents and deaths and injuries" from traffic accidents. Accordingly, the goal could be relatively easily met, as long as it could be demonstrated that some reductions had been accomplished.

While there were no details on resources or constraints, the legislation was very specific on *how* the improvements should be sought. Thus the National Traffic and Motor Vehicle Safety Act of 1966 required the inclusion of (1) driver performance, (2) pedestrian performance, (3) record systems, (4) accident investigations, (5) vehicle registration and inspection, (6) highway design and maintenance, (7) traffic controls, (8) vehicle codes and laws, (9) detection of potentially high accident locations, and (10) emergency services. The Act also called for (1) motor vehicle safety standards, (2) safety research and development, and (3) maintenance of a national driver register service.

A major thrust of the legislation was the setting of safety standards. Implicitly, if not explicitly, there were instructions to increase the safety of components, without stated restraints on how much safer components should be.

An evaluation of the policy represented by this legislation would consist of two parts: (1) whether the goals are being met and (2) whether the most effective methods are being used. These are two distinctly different questions, but both are relevant in considering the adequacy of an existing policy.

Goal Achievement

The goals of the 1966 Safety Acts are ambiguous. There are no instructions as to how priorities were to be decided, how much an improvement was

[7] See Selected Reading 4.2.

desired, the resources that could be used, or whether mobility could be reduced. Furthermore, "reduction in accidents," which is the goal, is not referenced to any datum. Therefore, whether a reduction was wanted in absolute terms, or a reduction in "what might have been," is not clear. Finally, standards were called for, but the quality level and the completeness of component standardization was not specified.

On the basis of a 1971 Report,[8] one could state categorically (1) that the goals *had* been achieved, or equally emphatically, (2) that the goals *had not* been achieved. There is no doubt that:

- There was a statistically significant reduction in the rate of loss; that is, the loss per vehicle miles of travel. Furthermore, numerous safety standards had been set.
- More persons were killed in highway accidents in 1972 than in 1966—similarly there was an increase in injuries, property damage, and the total number of accidents.

New legislation in the 1970s suggests that the Congress, the Executive Branch, and the crusaders were not satisfied with past results. One piece of legislation to improve automobile bumpers was primarily for "consumers' protection," since the Act could have little or no effect on fatalities. Another bill in 1973, provided for greater emphasis on removing adverse sections ("spot improvements") of highways.

Certainly the 1966 Acts implied that the highways were to be made "as safe as possible," and instructed the Executive Branch to accomplish this by setting minimum standards for the "safety" of components. Because limits were not set on mobility or on resources that could be expended, the constraints were left to the political resistance of individuals, industry, and governments.

Cost data for highway safety expenditures were available only in part in mid-1973. Absolute costs, or the costs of highway transportation that were directed solely to safety, cannot be rationally developed, since safety measures are intermixed with those for other operational characteristics.

To achieve the goal of being "as safe as possible," expenditures need to be "as productive as possible," assuming, that is, that there is some practical limit to resources. Therefore, one measure of a goal directed to achieving the maximum gain would be the relative effectiveness of actions that are taken. Since corrective measures can have short- and/or long-term benefits, both of these types of results should be evaluated.

A review of the effectiveness of highway safety expenditures is included

[8] See Selected Reading 4.10.

in the next section. But the adequacy of the time response can be addressed directly. If accidents are a function of the exposure (vehicle-miles) then a specified reduction of travel would have led to fewer accidents. Since such a reduction did not occur, then it follows that:

- The most rapid reduction of accidents was not achieved, in favor of retaining the growth in mobility.
- The policy (1966 to 1972) was to retain the growth in mobility.
- There is no requirement to reduce the total annual fatalities or accidents; rather it is to make the highways as "safe as possible" for the anticipated demands and within a "reasonable" time period.

Program Effectiveness

An evaluation of a program should include analysis of the benefits and costs of actions taken. Estimates are needed concerning which actions have been most effective and which have resulted in the least benefits. The results of this type of overall analysis by the government had not been made public by mid-1973, although some unpublished studies were available, as were reports on specific items.[9] The federal report on *Cumulative Regulatory Effects on the Cost of Automotive Transportation (RECAT)*[10] provided an economic benefit/cost analysis in 1972. The RECAT estimate of lives saved was:

1967	1,923
1968	3,047
1969	5,032
1970	9,080
Total	19,082

This was based on a reduction in fatalities, assuming (1) that the 1966 fatality rate of 5.7 fatalities per 100 million vehicle-miles had existed, and (2) that the savings in lives was the difference between the rate in 1966 and the rate in each subsequent year, multiplied by the vehicle-miles of travel for that year. For example, the fatality rate in 1967 was 5.50 (Figure 4.1),[11] which was 0.20 lower than for 1966. Travel in 1967 amounted to 965 billion vehicle-miles. Multiplying 9650 hundred-million vehicle-miles by 0.22 produces an estimate of 1923 lives saved in 1967.

[9] See Selected Reading 4.5.
[10] See Selected Reading 4.4.
[11] See Selected Reading 4.1.

ANNUAL BENEFITS

Looking first at the annual savings of lives which were estimated in the RECAT report, analyses should be made to determine what produced the benefit.

There were no actions taken under the 1966 Safety Acts that could have influenced the reduction in 1967. Some vehicle standards were applied, effective January 1, 1968, so that at least a half-year effect was present in 1968. Subsequently, 1½ years for 1969, and 2½ years for 1970 could be assumed for those standards that were made effective at the earliest date.

In 1967, approximately 2000 more miles of Interstate System became operational (this was the approximate rate of completion of the system during this period). At an average savings of 0.25 lives per mile per year, this would account for a savings of 500 lives. It is possible that the increased emphasis by the federal government on highway safety accounted for the additional 1400 lives saved in 1967.

It is also possible that 1966 was an abnormally bad year—that the variation was not statistically significant. In Figure 4.3, some hypothetical conditions are examined. Thus it may be that the fluctuations above and below a fatality rate of an assumed mean value (5.4) were normal variations that occurred for a relatively constant operating system. If this is so, estimates of lives saved that are based on the fatality rate of 1966 would not be valid. This assumption would account for the high value estimated for lives saved in 1967 and 1968—prior to effective action under the Safety Acts.

There is also a possibility that the system started to change in the early 1960s. Many persons born in the World War II "baby boom" became of driving age. This did not, however, produce a large increase in number of drivers.

Following 1967, it was predicted that seat belts and collapsible steering columns would produce significant benefits. The benefits from these changes were expected to be quite comparable, and would increase annually for 10 years or until all vehicles were equipped. Thereafter, there would be no further decrease in the rate, although there would be a continuing savings of life over conditions without these protective devices. This effect is also shown in Fig. 4.3, based on an average of 5.4 for the fatality rate.

These are maximum benefits because some redundancy exists. For example, the reduction of accidents on the Interstate reduces the number of fatalities that would have been estimated as being saved by seat belts and improved steering columns. Similarly, the Interstate System and seat belt estimates are influenced by the better steering columns.

The predictions for the combined effect of the Interstate System, seat belts, and the collapsible steering column would indicate a total reduction

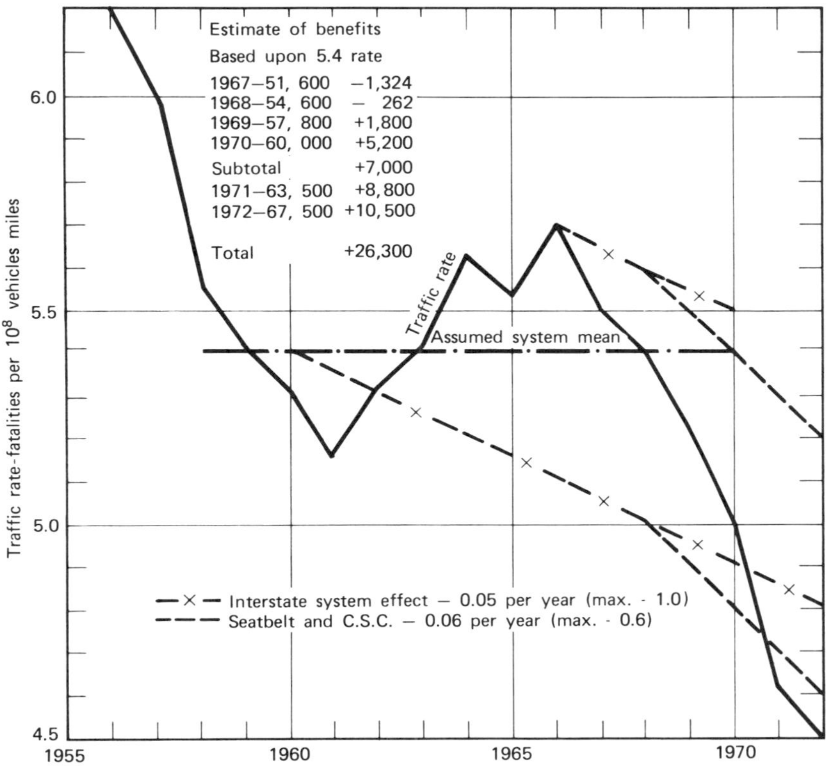

Figure 4.3. Estimates of benefits for 1967 to 1970 period.

of 0.8 in the fatality rate over the 1960 to 1970 period. The maximum estimate of the effect of these actions is to lower the fatality rate by 1.6.

Estimates of the reduction of the rate are more reliable than those for lives saved. In Figure 4.3, the estimate based upon an average fatality rate is 7000 for the period 1967 to 1970 inclusive, as compared to the RECAT estimate of 19,082. For 1972, the estimate is 26,300 as compared to the 46, 256 from the 1972 Traffic Safety Report.

The use of the average value of 5.4 for the fatality rate produces reasonable results. For example, the results for 1967 and 1968 are what might be anticipated during a program build-up. The effect of the Interstate System was relatively constant, and would not be expected to change following the 1966 Safety Act. The benefits from seat belts and improved steering column would accrue as vehicles became equipped. The final estimate of 4.6 (Figure 4.3) in 1972 compares favorably with the measured value of 4.5. Since the mean value was arbitrarily defined, Figure 4.3 demonstrates the range of the estimate of benefits that can be obtained, and may be a better estimate than one based on the 1966 fatality rate.

Insofar as annual savings are concerned, however, Figure 4.3 suggests that *if* the Interstate System, seat belts, and steering columns are performing as predicted, then it is quite possible that none of the other activities had a significant effect on the fatality rate. The relatively short time period of 2 to 4 years, plus the random variations, make such estimates quite difficult. Nonetheless, a steady reduction of the "average" fatality rate to a level of 3.8 due solely to the Interstate System, seat belts, and collapsible steering columns is not an unreasonable prediction.

Comparing this benefit with the 5-year goals of the example of the preceding section, and under 1979 traffic levels with normal growth, a rate of 3.8 would result in approximately 65,000 fatalities.

There is the possibility that other activities have had a substantial influence. Using the 1966 fatality rate, the reduction produced by the aforementioned three measures would have reduced the rate to 5.2 by 1972, as compared to the actual rate of 4.5. Therefore, the question is whether the high rate in 1966 was statistically significant. If so, what combination of measures could have produced the 0.7 reduction in rate? Based on the amount of travel in 1972, this would be a savings of nearly 9000 lives—or nearly eight times the number predicted for the combined benefits of the Interstate System, seat belts, and collapsible steering columns. In the absence of data to answer the question, it is only possible to conclude that the range of the benefits from the other activities may be very small to as large as 9000 lives for 1972.

A more detailed analysis can provide some guidance. For example, the percent of fatal accidents that occurred on wet pavements was as follows:

1966	16.7
1967	18.0
1968	16.3
1969	15.8
1970	16.2

The standards on windshield wipers, tire treads, slippery pavements, and similar actions directed to safety in wet weather produced very little change in the percentage of fatal accidents on wet pavements.

As a social problem, the fatality rate on a population basis is a more useful measure. In Figure 4.4, a perspective of this variable is provided. Since 1966, there has been a significant reduction in the rate of growth of the number of highway fatalities per 100,000 population. The level in 1973

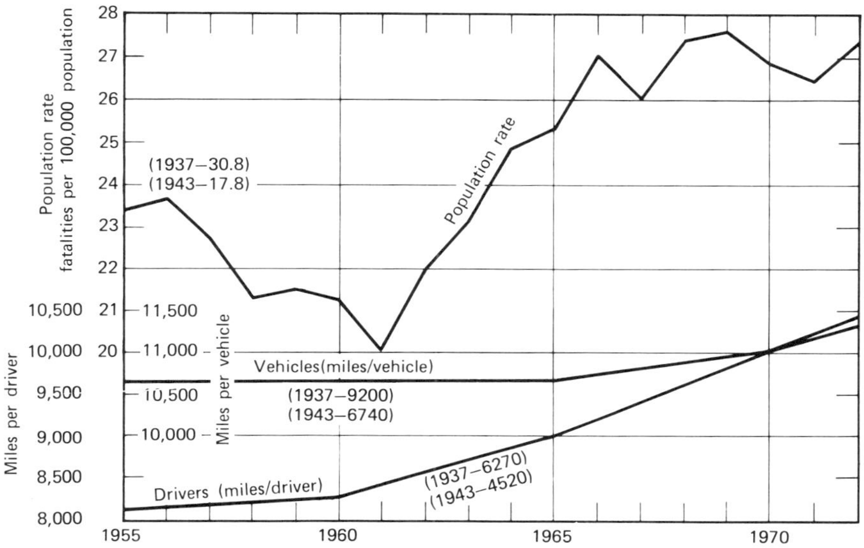

Figure 4.4. Change in population rate of fatalities and in vehicle and driver operations.

was approximately that of the preceding high in 1937 of 30.8 (see Figure 4.1). The reduction in growth rate occurred abruptly in 1967, and maintained the 1966 level thereafter.

Figure 4.4 shows the steady increase in miles per driver with a modest increase in vehicle-miles. The increase in both miles per driver and miles per vehicle was largest during the 1965 to 1972 period. Therefore, the leveling off of the population rate in light of the increased mobility (see also Figure 4.1 for increase in vehicle-miles) suggests that there have been improvements made since 1966. The level of 27 to 28 fatalities per 100,000 population occurred with four times as many vehicle miles of travel as compared to when this fatality level was reached in 1937.

In summary, relative to the goals of the 1966 Safety Acts, the following can be stated:

- There has not been a reduction in annual number of fatalities (or of injuries and accidents).
- There has been a reduction in the rate of growth of the annual fatality rate per 100,000 population.
- There has been a significant decrease in the fatality rate per 100 million vehicle-miles of travel.
- The amounts of improvements in the highway safety records since

1966 that can be traced to the Safety Acts are questionable, but there have been some benefits from actions taken in response to the Safety Acts.

- There is no basis on which to judge whether the progress has been maximum or sufficient, nor is there a generally accepted or reliable reference (or datum) upon which to measure benefits.

With reference to the goals of the example in the preceding section, the following can be stated:

- There is reason to doubt that the 2-year goal could be achieved under the existing program.
- Unless new actions are taken, there can be no objective basis for expecting that the 5-year goal could be reached. The only programs that were being discussed in mid-1973 that would have sufficient potential for reducing fatalities to 35,000 per year were the (1) reduction of use of alcohol and driving and (2) mandatory use of seat belts (or other passive restraint systems).
- Two research programs, one involving safer vehicles and the other concerned with electronic control systems, have the potential but would need to receive much greater priority and resources.
- Both of the aforementioned research programs have the potential for achieving the 10-year goal, but would require much higher priority and funding.

COST EFFECTIVENESS

As stated in the preceding sections, the greatest possible benefits are realized when the results per dollar spent (or other resource expended) are maximum. If the benefits that are desired are stated, then the total benefits need to be obtained even though the cost per unit of benefit becomes relatively high.

Table 4.8 provides a step-by-step procedure for conducting a cost effectiveness analysis for highway safety. Such studies require that either (1) quantitative goals are stated, or (2) a limit on resources is set. Given the first, the lowest cost to achieve the result can be estimated. Given the second, an estimate of the maximum benefits can be provided. When neither goals nor resources are provided, cost effectiveness can only provide a series of alternatives that show the cost of achieving various goals.

Cost effectiveness is one tool for the decision-maker. There are other factors to consider. For example, if school bus accidents, railroad grade crossing fatalities, or multivehicle catastrophes are of greater significance to society than other highway fatalities, then a premium price may need to be paid to achieve these specific goals. These special conditions, however, can be identified and included as constraints on the solution.

Table 4.9 provides a summary of cost effectiveness data, based on predicted benefits from published reports. The range in investments costs is from $4000 to $3.9 million per life saved. Based on annual costs, and if dollar values are assigned to travel time, the range is from zero (for highway construction) to $7.0 million per life saved.

When only earlier *predictions* are available (as in Table 4.9), the reliability of analyses of future programs is not too high. Nonetheless, a systematic and objective cost effectiveness analysis based on predictions can lead to better decisions than purely intuitive procedures.

After developing estimates of the benefits from individual measures, the effect of a combination of corrective measures needs to be prepared. Redundancy and timing are two important features of estimating the combined effect. Because of redundancy, measures to improve the highway, vehicle, driver, and emergency services need to be included in a single overall analysis. Otherwise, the benefits for individual measures are overestimated. Timing is important, because of both effects on redundancy and the ability to meet the desired objective within the time allotted.

Returning to the conditions of the example in preceding sections, it was not possible in mid-1973 to base the analysis on results from 1966 to 1972. Excluding the Interstate System, seat belts, and collapsible steering columns, the estimate of lives saved by highway safety measures ranged from near zero to 7000 lives for 1972. Expenditures were not available for state and local government programs, industry activities such as recall campaigns, and individual expenses on vehicle maintenance on safety features.

Insofar as meeting the 2-, 5-, and 10-year goals of the example, the standards that had been implemented through the end of 1971 had not been sufficient to produce fewer fatalities in 1972. There were a number of standards or changes of standards applied in 1971, 1972, and 1973 that should be evaluated to determine if sufficient benefits to achieve the 2-year goal were probable.

There was some evidence that the expenditures per life saved for the proposed standards would fall far short of achieving the 5-year goal. Consider the following costs (in excess of 1966 expenditures):

New vehicles (in 1976)	$4.5 to 7.5 billion
Vehicle maintenance	
($100/vehicle—25 per cent of vehicles	$2.5 billion
State and local government programs—1972 level	$2.0 billion

Table 4.9 Cost Effectiveness Data for Early Actions Taken under the 1966 Safety Acts

	Miles	Investment Cost Per Mile ($ Millions)	Total Investment Cost ($ Billions)	Lives Saved (Predicted) Per Mile Per Year	Lives Saved (Predicted) Per Mile 22 Years	Lives Saved (Predicted) Total 22 Years	Cost Per Life Saved ($ Millions) Investment Cost Basis FHWA	Cost Per Life Saved ($ Millions) Investment Cost Basis RECAT	Cost Per Life Saved ($ Millions) Annual Cost Basis[a]
FHWA Report (Selected Reading 4.5)									
Interstate									
Urban	6,780	5.0	34.5	0.45	10.0	67,800	0.50	0.90	2.7
Rural	33,420	1.2	40.5	0.18	4.2	144,000	0.28	0.26	1.2
Total	40,200[b]	1.9	75.0	0.24	5.2	211,800	0.37	0.46	1.3
ABC program—Per Year									
Urban	432	1.40	0.57	0.023	0.5	216	2.80		6.6
Rural	9,500	0.16	1.50	0.036	0.6	5,700	0.27		0.7
Total	9,932	0.21	2.08	0.036	0.6	5,916	0.35		0.9
TOPICS program									
Urban			2.214			561	3.9		6.6
Rural			0.116			30	3.9		7.0
Total			2.330			591	3.9		6.6
Highway safety improvement program (For 1971 Program)									
Urban				39[f]					0.33
Rural				181[f]					0.18
Total			0.300	220[f]		2,420[g]	0.11	0.13[d]	0.21

Railroad grade crossing[c]							
(One year)							
Structures		0.091	1.2[f]	240[h]	0.38		0.85
Protection devices		0.004	.5[f]	110[h]	0.004		0.09
Total		0.095	1.7[f]	350[h]	0.27		—
RECAT Report—Motor Vehicles							
(Selected Reading 4.4)							
Seat belts (25% use)	0.21[j]	2.1[k]	3200[l]	32,000[i]		0.07	0.12[e]
Collapsible steering column	0.18[j]	1.8[k]	2740[l]	32,000[i]		0.07	0.11
Interior padding	0.08[j]	0.8[k]	—	none		—	—
Head restraints	0.08[j]	0.8[k]	300[l]	3,900[i]		0.27	0.47
Dual braking	0.10[j]	1.0[k]	410[l]	4,900[i]		0.26	0.41

Source. From Selected Reading 4.2.

[a] Annual cost at 10% for 20 years, except as noted.

[b] Toll Roads—2,300 miles.

[c] From Selected Reading 4.5 (August, 1972): Total cost—$750 million
Total benefits—10,000 lives saved
Cost/life saved—$75,000

[d] 20-year life, based upon investment cost.

[e] RECAT Costs based upon 10% for 10 years.

[f] Total Investment Cost per year.

[g] Total Investment Cost for 11 years.

[h] Total Investment Cost for 20 years.

[i] Total lives saved in 10 years.

[j] Cost per year.

[k] Cost for 10 years.

[l] No. of vehicles per life saved per year.

Note. If the benefit/cost ratio for an improvement is greater than 1.0 over the life of the improvement, exclusive of dollar savings placed upon reducing fatalities, then on an annual cost basis there is no cost for saving a life. If the cost per life saved is greater than savings placed upon reducing fatalities, the program can be justified on humanistic but not economic grounds. Values for cost per life saved were computed from data on benefits and costs contained in Selected Reading 4.4 and 4.5.

Objective appraisals of the various standards would produce estimates of the ability to meet either the 5- or the 10-year goals. On the basis of predicted benefits, however, there is reason to believe that a more cost effective program is needed, particularly if mobility is not to be drastically reduced.*

On the basis of 1973 conditions, 50,000 lives would need to be saved each year, at an annual cost of $30 billion. This would require an overall cost effectiveness of $600,000 per life saved. If mobility is retained, the cost increases. If economically productive techniques are used, part of the cost is offset.

STATE AND LOCAL GOVERNMENTS

Policy development at state and local government levels differs primarily from that at the national level in the constraints that are imposed. These constraints may be specific limits that are set by the federal government, or they may be factors that are too pervasive for local governments to control. An example of the latter would be control of vehicle characteristics; that is, it is usually impractical, illegal, or not wise politically to establish unique requirements for vehicles used within a state or an urban area.

Highway safety offers a particularly good example of the limitations that exist for policy development at local levels. These are discussed in the following section.

Goal Setting

Needs are more easily perceived by state and local governments because of the relatively small areas and populations that need to be considered. Furthermore, the government is much closer physically to the governed. Local press, public interest groups, industry, and individuals provide the political system with a continuous description of the needs.

But in addition, the local community's share of meeting national needs is also inherited. While national requirements are usually accompanied by financial assistance, it is also usual for the local levels to provide matching funds. On occasion, most of the funding must come from local sources, and so it is with highway safety. For example, in the early 1970s, it was esti-

* The reduction in one year—1974—was several times greater than the reduction achieved during the preceding 6-year history of the Highway Safety Acts. This suggests that the several billion-dollar expenditure for highway safety was allocated to a program with a relatively low cost effectiveness.

mated that state and local governments would spend $2 billion per year, at a time when the *total* national budget for highway safety was less than $400 million (of which less than half was for matching funds for local programs).

The local incentive to respond to national priorities for highway safety was largely to avoid penalties. If satisfactory programs were not developed in the specified priority areas, the states were faced with a loss of 10 per cent of their federal highway funds. Therefore, local policy development for highway safety is influenced greatly by federal policies.

The perception of need in highway safety becomes confused then at the local level. The direct losses in fatalities, injuries, and accidents become obscured by federal program requirements which are for the development of satisfactory standards. The sixteen standards listed in Table 4.10 were mandatory in the early 1970s. There were no specific requirements for reducing the number or severity of highway accidents, nor were there maximum or minimum limits set on funding. There were also no minimum or maximum conditions that the standards had to meet—only that they needed to be "adequate."

Table 4.10 List of Highway Safety Standards[a]

No. 1.	Periodic motor vehicle inspection
No. 2.	Motor vehicle registration
No. 3.	Motorcycle safety
No. 4.	Driver education
No. 5.	Driver licensing
No. 6.	Codes and laws
No. 7.	Traffic courts
No. 8.	Alcohol in relation to highway safety
No. 9.	Identification and surveillance of accident locations
No. 10.	Traffic records
No. 11.	Emergency medical services
No. 12.	Highway design, construction, and maintenance
No. 13.	Traffic engineering services
No. 14.	Pedestrian safety
No. 15.	Police traffic services
No. 16.	Debris hazard control and cleanup

[a] These are highway standards that must be met by local governments, as established by the National Highway Traffic Safety Administration in accordance with the provisions of the Highway Safety Act of 1966. Regulations call for an adequate safety program in each of the areas covered by a standard. There is an additional set of standards for motor vehicles, in accordance with the National Traffic and Motor Vehicle Safety Act of 1966 (see Selected Reading 4.10).

Certainly the fulfilling of national standards does not deal directly with the local perception of needs in highway safety. Furthermore, some of the standards can be more effective than others, depending in part on local conditions. To develop the most effective programs, state and local governments should focus on reductions in some specific consequence of highway accidents, and should establish the level of resources which are to be used and any constraints to be imposed. The local goals also need to be explicit as to how national standards will be treated.

By setting goals, such as a reduction in annual fatalities, a sound basis is established for policy decisions on how local resources should be used. Furthermore, a sound basis can be established for the development of standards in response to national requirements.

Evaluation of Alternatives

The effect of national programs must first be estimated before an evaluation of local alternatives can be made. In particular, the effect of (1) vehicle standards and (2) highway construction should be evaluated.

The benefits from new vehicle standards can be estimated for local conditions and for the timetable under which these benefits accrue. The administrative requirements for local governments need to be established. For new vehicles, these are nominal. For used vehicles, however, local policies control the requirements and the costs both for local governments and for individuals. Thus tire maintenance and vehicle inspection require local police action before any program can become operational. These alternatives need to be compared with other alternatives for achieving local goals.

Highway construction, within limits, can also be controlled to produce the greatest contribution to local goals. Even though federal aid is involved, local governments have the option of selecting which routes are improved and how this improvement is made. The limitations that are imposed are (1) the amount of funds available and (2) the system on which the funds can be expended.

A major alternative exists for the allocation of highway funds. The conventional approach is to select improvements primarily on the basis of benefit/cost ratios; that is, priority is given to the project with the largest economic benefit/cost ratio. If, however, benefits are expressed as safety improvements, the projects with the greatest return in safety per dollar expended could be identified. Accordingly, priorities could be assigned so as to achieve the greatest safety gains.

After accounting for the benefits that can be anticipated from federal action, the evaluation of alternatives for local policy actions can proceed. A much narrower set of alternatives exist. For example, modification of new vehicles is a federal responsibility.

On the other hand, local policy can give much more critical attention to the requirements that call for local government funding. Driver education, policing, and control of alcohol and driving produce specific financial and manpower needs. The social, economic, and political feasibility of such programs can be more realistically evaluated. Cost effectiveness analyses can produce the program with the greatest benefits per dollar expended.

Local policy analyses are also able to deal more directly with expenditures from the private sector, particularly from individuals. Private costs for driver education, vehicle inspection and maintenance, restrictions on driver licensing, and alcohol control measures can be accurately appraised, both in financial terms and in needs for transportation.

SELECTED READINGS

4.1 *Accident Facts,* National Safety Council, Chicago, Illinois, Annual Editions.

4.2 Baker, R. F., "A Critical Analysis of Federal Highway Safety Policy," *Accident Analysis and Prevention.* New York: Pergamon Press, December 1973.

4.3 Baker, R. F., *The Highway Risk Problem: Policy Issues In Highway Safety,* New York: Wiley-Interscience, 1971.

4.4 *Cumulative Regulatory Effects on the Cost of Automotive Transportation (RECAT),* Final Report, Office of Science and Technology, Washington, D.C., February 28, 1972.

4.5 *Evaluation of Costs and Effectiveness of Major Federal-Aid Highway Programs,* Special Analytical Study, Federal Highway Administration, July 1970.

4.6 *National Commission on Product Safety, Final Report,* National Commission on Product Safety, Washington, D.C., 1970.

4.7 "Occupational Safety and Health," *Standardization News,* American Society for Testing and Materials, Vol. 2, No. 1, 1974.

4.8 *Perspective on Benefit-Risk Decision Making,* Report by the Committee on Public Engineering Policy, National Academy of Engineering, Washington, D.C., 1972.

4.9 "Product Liability," *Standardization News,* American Society for Testing and Materials, Vol. 1, No. 3, 1973.

4.10 *Safety '71, An Activities Report,* National Highway Traffic Safety Administration, August 1972.

4.11 *Safety '72, An Activities Report,* National Highway Traffic Safety Administration, October 1973.

4.12 "Standards for Consumers," *Standardization News,* American Society for Testing and Materials, Vol. 1, No. 10, 1973.

4.13 Starr, Chauncey, *Social Benefit versus Technological Risk,* University of California at Los Angeles, 1968.

CHAPTER 5

POLICY DEVELOPMENT FOR A NEW TECHNOLOGY—UNDERGROUND CONSTRUCTION

A very frequently encountered public policy problem in the midtwentieth century was whether to invest in the development of a new technology. This question arose in many fields—energy, health, and space to mention a few. For one of the more glamorous of these, the federal government made a major commitment to space technology, and resisted (through 1974) technology for the supersonic transport.

Not all of the pressures were for glamorous innovations, and they included such ideas as electronic communication and control systems for highway vehicles and a mile-deep tubular, gravity transport system. Innumerable other ideas were advanced for applying advanced technology to the solution of social programs.

Pressures to fund new technology came in the form of research proposals from many segments of the field of science, technology, and engineering. A common characteristic of these proposals was that clear benefits to society were inevitable if the technology were to become operational.

Not available to the policy-maker, however, was an objective measure, or even the capability to measure, the relative significance of the new technology. It was not possible for the policy-maker to compare one with another, or to know how the new technology would increase or decrease the most serious problems facing society. The funding needs were far in excess of the resources available.

The following case study is directed to this dilemma. It is concerned with the question of resource allocation to a new technology—in this case, underground construction. As it becomes apparent, the study was handicapped by the absence of (1) policy statements on national goals and priorities, (2) a consensus on the relative severity of social problems, and (3) a generally accepted measure of individual or collective well-being. Nonetheless, resources were being allocated, and implicitly (or explicitly) these questions were being answered.

While the case study approaches a problem that requires a nationwide approach, the principles and thought processes can be applied to regional or local problems that involve lower order changes in technology.

This type of policy development can be undertaken by either a protagonist, antagonist, or neutral group that is concerned with the problem of what type of impact a new technology has on the social, economic, and environmental characteristics of a community. If explicit policy goals for life qualities are available, the following type of case study would be one phase of a larger study. If such goals are not available, a study of the following type would be useful as a basis for comparison with other ways for allocating resources.

The example is based on a policy development study that was summarized in a report released by the American Society of Civil Engineers in 1972.[1] This study was conducted by the Underground Construction Research Council (UCRC) under the sponsorship of the National Science Foundation, the Engineering Foundation, and contributions from the private sector. This chapter is an extensive summary and critique of the UCRC study as a typical example of—and one model for—policy development in a major technology. The study, which took 6 months to complete at a cost of $100,000, was conducted by an interdisciplinary team of seven professionals, including two of the authors of this book.

The major issue addressed by the Underground Construction Research Council was whether a substantial increase in the use of underground space should become a national goal. If so, a quantitative statement of the goal, the resources, the timetable, and other constraints would be required. Furthermore, because underground construction was not a major activity of either the private or the public sector, the problem was considered to require national as opposed to local emphases.

GENERAL PLAN

The objective of the study was stated as follows: "to determine the total consequences of increased use of underground space, including alternative requirements for research and development." The following guidelines were established:

- A decision to make increased use of underground space was a political, not a technical one.
- An objective appraisal would be needed by decision-makers.
- The technical role was to implement the policy decision.
- A policy commitment on program requirements, resources, timetables, and type of change to be made was needed for science and technology to make their greatest contribution.
- Policy development studies were new, data were limited, and some

[1] *The Use of Underground Space to Achieve National Goals: Report of a Program for Improving the Effectiveness of Underground Construction*; prepared by Robert F. Baker and six others, including Richard M. Michaels; conducted by the Underground Construction Research Council of the American Society of Civil Engineers (New York: The Society, 1972). We have made generous use of tables and figures from the UCRC report in this chapter, with occasional adaptations in title to fit their context here. These and all direct quotations are used with the permission of the Society. In all subsequent footnotes in this chapter, whether in text, table, or figures, *ibid.* refers to the report.

research and development in study methodology would therefore be involved.

The plan for developing alternative program requirements is shown in Figure 5.1. A unique feature of this type of evaluation is that the specific social goals (Figures 2.4 and 2.5) were not established. Rather, the impact on the social, economic, and environmental characteristics that could be social goals were to be estimated.

Because the statement of explicit goals was not an aim of the study, it could not be used by policy-makers as a substitute for decision-making. Policy-makers would need to decide on the social goals, analyze the many alternatives suggested by the study, and then choose a goal and a possible way to meet that goal. This would, of course, include the possibility of selecting means, or even goals, not discussed in the report. For example, a

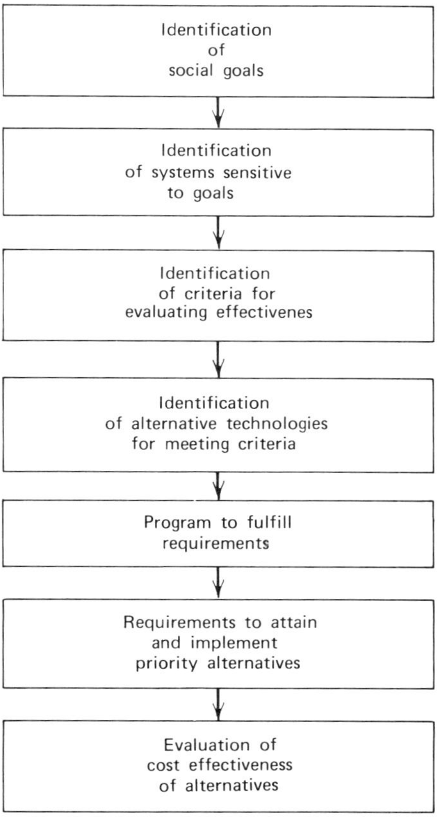

Figure 5.1. General plan for developing Program Requirements. *Source. UCRC Report*, p. 6.

report prepared by technologists can estimate the impact of improved highways (or change in other means of transportation, or in use of water resources, or in generation of electrical energy). But it cannot decide on specific social goals. Policy-makers must study the alternatives, and then decide what social goals to meet and how to go about it, basing their decision on the sound technical information provided by such a report.

The UCRC example is comparable to other functional analyses wherein the benefits from one specific activity or group of similar activities are analyzed.

So-called analyses can also become merely advocacy studies or arguments for a particular solution. Advocacy studies or arguments for a particular solution sometimes masquerade as "functional analyses." Therefore, a good policy analysis needs to be objective, the assumptions explicitly stated, and the study as comprehensive as possible. As noted earlier, an advocacy report is a political, not a technical strategy, and is not the most useful type of study on which to base a policy decision. As a final step in the UCRC study, the feasibility of alternatives was considered from a technical, social, economic, and political view.

The foreword of the ASCE report on underground space also includes a political recommendation; that is, a particular solution is recommended by UCRC on the basis of recommendations of the project staff. Usually, recommendations are a part of a policy development study. To avoid misleading the decision-maker, however, the assumptions on which the recommendations are based are included.

In the following example, the primary objective is to establish a goal in a manner that is sufficiently precise to permit the most effective implementation. Accordingly, the question of *how* any change should be accomplished is examined only to determine general technological feasibility. With reference to Figure 3.1 then the primary objective is "Goal Definition"—step 1, which is accomplished by a relatively superficial conduct of steps 2, 3, 4, and 5 of Figure 3.1.

IDENTIFICATION OF SOCIAL GOALS AND CRITERIA

Instead of setting social goals, the study listed and discussed the following needs which were perceived as having high priority by the general public for social and environmental systems:

- Reduction of poverty
- Reduced unemployment
- Ethnic and racial equality

- Improved mental health
- Improved physical health
- Increased safety for persons and property
- Elimination of urban decay
- Improved education
- Relief of urban congestion
- Relief from residential crowding
- Improved housing quality
- Decreased air and water pollution

In addition, as measures of social qualities, the study presented the following criteria for social goals:

- Opportunity time
- Public health and safety
- Job opportunity
- Esthetic qualities
- Environmental quality
- Shelter (quality, availability, and density)
- Mental health
- Ethnic quality
- Conservation of natural resources

Each of these is discussed in more detail in the following section on evaluation.

Economic criteria were selected based on the net dollar effect of locating facilities in the underground rather than on the surface. The economic criteria, discussed in more detail in a later section, were as follows:

- Direct cost
- Time
- Land
- Energy
- Pollution
- Safety
- Reliability
- Material resources

The analysis in the report proceeded on the assumption that the preceding list of social and economic characteristics were the relevant factors to the policy-makers. The ultimate decision—that is, the extent to which resources would or could be allocated to underground construction—would depend on the proposed construction's impact upon specific social conditions, or rather, upon the policy-makers' understanding of its possible impact. Those who prepared the UCRC report also assumed that similar analyses of other alternatives would need to be made before the policy-maker could understand fully the relative effectiveness of resource allocation. Although alternatives to underground construction were considered in the report, as discussed in a later section, the *total set* of societal alternatives was not considered.

SYSTEMS SENSITIVE TO GOALS

Underground construction can be expected to have a wide range of impacts on human existence, depending on the type of facility and activity that is housed in subsurface space. Certainly the impact of a highway tunnel, for example, is different from that of an underground warehouse or power plant.

Life-Support Functions

To develop insight concerning the different benefits and adverse effects of the use of underground space, the report classified life-support functions into the following eight groups:

- Residential shelter systems
- Commercial, industrial production structures
- Water resource systems
- Liquid waste management
- Solid waste disposal
- Transportation
- Communication
- Energy

The report described each of these functions as summarized below.

1. *Residential shelter systems.* These systems included the structures and related appurtenances required to support individual and family life. Individual use of space, and the environment that is needed, were included in the analyses.
2. *Commercial, industrial production structures.* Organized production, communication, and exchange activities require a specialized ar-

rangement of facilities that produces effective and efficient operations to achieve a variety of missions. Both structural form and services are of interest. This group includes factories, warehouses, multistoried commercial structures, high-rise office structures, auditoriums, and other facilities for recreational and cultural activities.

3. *Water resource systems.* Water is one of the most basic life-support requirements. To provide water in sufficient quantity and quality for its many different purposes, a complex system is required for locating, transporting, treating, and distributing large quantities. Many parts of this system were already located in the underground in the early 1970s. Potential benefits were seen to lie in additional use of underground space and in improved construction technology for future installations.

4. *Liquid waste management.* A wide range of society's liquid waste products must be collected, treated, and disposed of. This activity must be accomplished in a way that protects the producer of the waste as well as the environment into which waste products are introduced. In the early 1970s, much of the collection of liquid waste was made possible by a network of underground facilities. Potential benefits were seen to exist in greater use of underground space for treatment purposes and in improved construction technology.

5. *Solid waste management.* This category included the waste products which were not being treated by disposal through waste water treatment. The collection, conversion, and/or disposal has unique system requirements. In the early 1970s, solid waste was a labor intensive activity, with large potential benefits from collection and disposal through use of underground space.

6. *Transportation.* The movement of persons and goods is another essential part of an organized society. For most social and economic activity, transportation by highway, air, rail, water, or pipeline is needed. For surface movements, rail and highways dominated transportation in the early 1970s. Use of the underground was rare, except in mountainous areas or in urban locations where surface facilities would have been disruptive of community life.

7. *Communication.* Included in this category is the exchange of information between individuals (or functions). This important attribute or requirement of a modern society involves the transfer of information by means of interpersonal exchange, paper exchange, or electronic transmission. The latter two require technology that could be influenced by increased use of the underground.

8. *Energy.* Large sources of energy are required to support life, particularly in a complex, urban society. The energy must first be produced in usable form, and then distributed in an efficient, reliable manner.

Electrical energy was the largest energy form in use in the 1970s, and secondarily liquid and gaseous fluids. Requirements to meet demands for energy call for power plants and distribution systems, all of which could be located in underground space.

Technologies Upon Which Life-Support Functions Depend

The persons who prepared the UCRC report made a study of the technologies required to support the preceding eight functions. The major components, listed in Table 5.1, are analyzed in terms of their sensitivity to

Table 5.1 Operational Components of Selected Technologies

Transportation

1. Highways
   a. Roads and pavements
   b. Vehicle
   c. Traffic control
2. Airports
   a. Layout
   b. Construction
   c. Terminal development
3. Railroads
   a. Track construction
   b. Equipment
   c. Control
4. Pipelines
   a. Petroleum
   b. Construction

Water resources

1. Planning
2. Water supply
3. Hydrology
4. Hydraulic structures
5. Hydraulic machinery
6. Drainage
7. Flood control
8. Irrigation

Solid waste disposal

1. Types of refuse
2. Disposal methods

space requirements including several hundred components. From these, the effects of technology are analyzed as discussed in the following section.

EVALUATION OF EFFECTIVENESS

The location of the life-support functions produces a series of predictable consequences. For those systems located in the underground, problems arise from technological limitations in design, construction, maintenance, and operations.

For surface installations, the design must account for environmental as well as operational forces and variations. In particular, climatic conditions produce specific resource requirements which are markedly different from

Table 5.1 Continued

Electrical energy
1. Energy sources
2. Conversion systems
3. Generators
4. Transformers
5. Switching systems
6. Control devices
7. Transmission
8. Distribution
9. Operations
Production systems—industrial structures
1. Use
2. Financing
3. Design
4. Construction
5. Maintenance
Communications
1. Information sources
2. Transducers
3. Transmitters
4. Transmission media
5. Switching
6. Amplifiers—repeaters
7. Receivers
8. Displays

those for underground space. Geographical and ecological factors are also present and provide severe limits on land use. In fact, these factors frequently define the accessibility to and development of a region.

The UCRC report analyzed each of the eight life-support functions to determine all apparent benefits from locating underground. Table 5.2 shows a listing of sets of these benefits. From these sets, the UCRC analysts went on to make estimates of the improvements that would result from locating the function in underground as opposed to surface space.

Economic Benefits

To determine the economic consequences of placing all life-support functions in the underground rather than on the surface, and to make this determination for the entire nation, a very broad and general approach was necessary. In addition, under limitations of time and data, very approximate results were all that could be developed. The UCRC study produced "scouting approximations"; that is, they were sufficiently accurate to produce a reasonably reliable estimate of the order of magnitude of the economic consequences.

The plan for developing economic consequences consisted of the following.

- Adoption of simplifying assumptions
- Selection of criteria
- Estimate of economic benefits

Three major assumptions were used to reduce the complexity of the analysis of economic consequences. These assumptions led to conservative results. More accurate estimates could be developed by adjustments that account for the effect of the assumption.

First, *the report assumed that the investment costs were equal*; that is, that the construction costs for subsurface construction were the same as for surface construction. This was not the case in the early 1970s. In fact, the ratio of construction costs, exclusive of social effects, ranged from 3:1 to higher than 10:1 (subsurface:surface). This assumption was originally conceived as a way to simplify the first approximations. As seen in subsequent paragraphs, however, economic benefits based on this assumption provided a sound basis for the evaluation of alternative policies. The assumption provided an interesting political perspective; that is, it provided an estimate of the extra costs that were required in order to live and operate the life-support functions on the surface rather than in subsurface

Table 5.2 Effects of Underground Construction Technology

Cost Category	Direct Effect, or Effect Unique to System	Indirect Effect on Other Systems
Transportation		
Economic	Surface land placed on tax rolls Reduced design requirement for road bed for environmental variations Reduced maintenance for roadbed right-of-way and traffic control devices due to reduced environmental interaction	
Time	Travel time becomes independent of weather conditions	
Land Urban Rural	Acquisition cost eliminated	Land surface available for building or open space use
Energy	Increased cost for lighting of roadway	
Pollution controls	Noise pollution eliminated Air pollution control on environment rather than individual vehicles	
Safety	Decreased accidents due to elimination of wind, snow, ice, water	
Reliability	Slow-downs due to snow and ice are eliminated	
Resource expenditures	Pavement reduction due to reduced environmental interaction	Slightly larger transportation tube could also accommodate: 1. electrical power distribution 2. waste water systems 3. water resource systems 4. communications systems

Table 5.2 (Continued)

Cost Category	Direct Effect, or Effect Unique to System	Indirect Effect on Other Systems
Transportation (cont.)		
Social	Residential, industrial, and commercial relocation eliminated	Increased stability of residential areas
Water Resources		
Economic	Reduced labor cost for construction of pipelines Reduced disruption during construction Elimination of trenching operation	Elimination of patching of highways due to installation of new lines
Time	—	
Land	Reduced restrictions on surface use of easements	
Energy	—	
Pollution controls	Reduced pollution due to controlled entry into reservoirs	
Safety	Elimination of hazards involved in trenching process	
Reliability	Reduced failures due to increased quality control of construction	
Resource expenditures	Reduction or elimination of select material for bedding pipelines Elimination of replacing surface material destroyed during construction (pavement, sod, etc.) Savings in water lost due to evaporation in open reservoirs	
Social	—	

Table 5.2 (Continued)

Cost Category	Direct Effect, or Effect Unique to System	Indirect Effect on Other Systems
Waste Water Control		
Economic	Reduction in labor cost for construction of pipelines Reduced disruption during construction Elimination of trenching operations and consequent dewatering problems	Elimination of patching of highways, etc. when new lines are installed
Time	—	
Land	Reduction in or elimination of restrictions on surface use of easement	
Energy	—	
Pollution controls	Contained treatment of odor from treatment plants possible	Relaxation of restraints on location of other systems relative to treatment plants
Safety	Elimination of hazards involved in trenching operations	
Reliability	Reduced failures of pipelines due to increased quality of construction	
Resource expenditures	Reduction or elimination of select material for bedding pipelines Elimination of replacing surface material destroyed during construction of pipelines (pavement, sod, etc.)	
Social	—	

Table 5.2 (Continued)

Cost Category	Direct Effect, or Effect Unique to System	Indirect Effect on Other Systems
Shelter		
Economic	Reduced costs due to reduced environmental interaction 1. reduced structural design requirements (wind, snow and ice) 2. reduced atmospheric control system requirements 3. reduced storm water removal costs—guttering and storm drains Reduced exterior maintenance costs due to weathering—painting, rust prevention	
Time	—	
Land	Elimination of land requirement	Allows higher density—open space may now be above residential structures
Energy	Reduction in total energy required for environmental control—heating, lighting, and air conditioning	Reduces water pressure requirements in distribution lines
Pollution controls	Reduced water pollution during construction (erosion and consequent suspended solids in water) and operation (more natural drainage without external pollutants)	Reduces visual and rural conflict with surrounding space uses
Safety	—	
Reliability	Reduced susceptibility to environment damage (water, wind, snow and ice, and fire)	

Table 5.2 (Continued)

Cost Category	Direct Effect, or Effect Unique to System	Indirect Effect on Other Systems
Shelter (cont.)		
Resource expenditures	Savings in insulation material, architectural exterior material, and traditional roofing materials	
Social	Consider effect of visual isolation from surroundings	
Electrical Energy Distribution		
Economic	Reduced maintenance (due to weathering) Elimination of overload systems designed to cope with failures caused by adverse enviromental interactions Structural design for environment reduced Reduced construction cost—specialized equipment for above ground work eliminated	
Time	—	
Land	Elimination of planning constraints	
Energy	—	
Pollution controls	Elimination of visual pollution of unsightly poles and towers	
Safety	Reduced accidents from high work Reduced accidents involving accidental contact between electrical and other systems	

Table 5.2 (Continued)

Cost Category	Direct Effect, or Effect Unique to System	Indirect Effect on Other Systems
Electrical Energy Distribution (cont.)		
Reliability	Improved reliability by eliminating failures caused by wind, snow and ice, and fire	Improved reliability of other systems dependent on electrical power
Resource expenditures	Elimination of tower and pole structures Elimination or reduction in overload systems	
Social	Reduction in conflict over right-of-way location decisions	Esthetically improved environment
Production Systems and Industrial Structures		
Economic	Reduced storm water control investment Reduced structural design requirements (wind, snow, and ice) Reduced heating costs Reduced construction costs (due to sub-system eliminations) Reduced maintenance (painting, cleaning, rusting)	
Time	—	
Land Urban Rural	Land requirement eliminated (post construction)	Surface space required for access if transportation system is on surface
Energy	Elimination of A/C requirement Reduction of heating costs	
Pollution controls	Unsightly buildings eliminated Noise pollution reduced	Reduced storm water input to storm sewer system

Table 5.2 (Continued)

Cost Categtory	Direct Effect, or Effect Unique to System	Indirect Effect on Other Systems
Production Systems and Industrial Structures (cont.)		
Safety	Increased construction safety due to reduced environmental interaction	
Reliability	—	
Resource expenditures	Excavated material available for other use Reduced material costs	Reduced electrical (or other source) energy required
Social	—	

Source. Ibid., pp. 116–117.

space. For many structures, total operating costs range as high as 10 or more times the annualized investment cost. Therefore, the significance of the equal construction cost assumption varies. The importance of this issue in an example of policy analysis, however, is the simplicity which is introduced by the assumption. Furthermore, analyses can be made in a way that permit more accurate appraisals by various subsequent assumptions based upon differential construction costs.

Second, *the report assumed that no existing policies would be changed relative to the quantity or quality of service*; that is, it was assumed that policies relative to highway travel, energy needs, water pollution, and such would not be influenced by locating facilities in underground space. This, in effect, compares operations on the surface to those same operations located underground.

Third, *the report assumed that there would be no interactions as a result of greater use of underground space.* This assumption, in the overall, provides a conservative estimate of economic benefits. For example, if highways were to be located in tunnels, then more water supply, waste collection, and energy and communication lines could be integrated into the corridor at less cost. Also, the effect of changing land values was not considered in the report. Under early 1970 conditions, land costs were a large part of construction costs, and with more facilities underground, land values would change.

There was also no account taken of the influence of underground locations on space use, or on technology of life-support systems. Certainly the relative location of life-support systems would take advantage of the pattern of space use. For example, very high speed transport would be more likely to evolve under tunnel conditions that restrict or limit interference or interactions with other activities.

For policy analyses, the foregoing types of assumptions lead to sufficiently accurate results for policy makers, and can be conducted with data available in the 1970s and within tight timetables. But it is extremely important to state the assumptions in a precise manner to avoid a misconception of the accuracy of the results. When possible, the assumptions should permit conduct of subsequent analyses that increase the accuracy of the first approximations, rather than require a completely new approach.

SELECTION OF CRITERIA

The economic analyses for the UCRC study identified eight cost functions that could be significantly influenced by surface versus subsurface location. These, listed in a preceding section, are discussed in the following paragraphs.

Direct cost included all *operating* costs that could be significantly influenced by the difference in location. Differences in construction costs were *not* included, in accordance with the aforementioned assumptions.

Time savings can provide economic benefits if economic productivity results. Two classes of time savings were identified in the study. One of these is the time lost in surface transportation because of environmental factors thereby decreasing working time. The second is the time lost in production functions due to environmental damage such as from fire and flood, or indirectly due to failure of supporting services such as energy or water supply.

The UCRC estimate included only those time losses traceable to the production functions. Certain other costs were not included because of uncertainties or lack of readily available data. For example, poor travel (environmental) conditions lead to lower retail sales.

Land costs were included on the basis of the surface space that would be released for other purposes. Since these costs were included as an operational savings, they were not included as a savings in construction costs.

Energy savings are largely the result of reductions in heating, ventilating, and air-conditioning. Structures that are located at least 15 to 30 feet below the ground require much less energy to maintain at comfortable temperatures. To provide for a temperature rise of 10 to 20 degrees requires much less energy than does providing for a 50 to 100 degree variation. Lighting requirements might be offsetting. However, lighting is a very small part of the energy requirements. Furthermore, surface structures in the mid-twentieth century require almost continuous lighting.

Pollution of the environment results to a major extent from the operation or construction of life support systems. Air pollution can largely be traced to transportation and production systems. Water and solid waste pollution stem from the technologies of production and/or the methods of disposal. The potential savings are large through greater use of underground space. Trapping the polluted air and releasing it to the surface after treatment could be much less expensive than the conversion and maintenance of surface systems. Similarly, the collection and treatment of water and solid waste could be more efficiently and reliably accomplished in underground space.

Increased *safety* offers potential for both economic and social benefits. Risks from flood, fire, wind, and civil system operations (highways, for example) could be greatly reduced in a controlled environment.

Reliability of some life-support systems can have a significant economic effect. Interruption of service is costly to producers and consumers. Electricity, communications, transportation, and water supply are systems that require high reliability to avoid economic losses.

Material resources are a major cost in construction of facilities for life-support systems. The shift of facilities to underground space would significantly reduce the quantity and quality requirements for construction materials. Roofing, guttering, insulation, and wind load structural requirements would be eliminated. Many hydraulic structures in transportation would also be eliminated.

ESTIMATE OF ECONOMIC BENEFITS

Based on the preceding assumptions and for the selected criteria, the UCRC analysts prepared a simple table of cost savings per life-support function (Table 5.3). Since the units are dollars, the columns can be summed in both directions, leading to an estimate of total cost savings. The savings were prepared on an annual cost basis, leading to the total economic savings of $58.6 billion that is shown.

The cost estimates were described as "crude at best." The lack of data, lack of time (or accessibility) to obtain available data, and absence of basic relationships reduced the accuracy. In general, the cost savings were considered to be on the conservative side; that is, the cost savings would probably be larger than estimated. The margin of error of the overall estimates was considered to be in the range of −50 to +150 per cent.

From a study of Table 5.3, certain major conclusions can be drawn. For example, direct costs and material resources account for nearly 40 per cent

Table 5.3 Social and Economic Benefits Accruing from Transferring Civil Works Functions to Subsurface Space

Benefit Category	Shelter	Production Systems and Industrial Structures	Water Resources	Liquid Waste	Solid Waste	Trans-portation	Commun-ications	Energy Distri-bution	Total
A. Economic benefits ($ billions)									
1. Direct costs	2.8	3.5	1.0	1.0	2.0	2.3	0.5	1.2	14.3
2. Time	0	4.0	0	0	0	8.0	0	0	12.0
3. Land	1.0	0.3	0	0	0	1.0	0.1	0.1	2.5
4. Energy	3.8	3.8	0	0	0.1	0	0	0	7.7
5. Pollution	0	0	0	1.3	0.1	4.0	0	0	5.4
6. Safety	3.0	0	0	0	0	3.1	0	0.1	6.2
7. Reliability	0	0	0.1	0	0	0.2	0.1	1.0	1.4
8. Material resources	3.0	2.5	0.8	0	0.8	1.0	0.5	0.5	9.1
Total	13.6	14.1	1.9	2.3	3.0	19.6	1.2	2.9	58.6
B. Social and environmental benefits									
1. Opportunity time (billions of man-hr)	10.0	6.0 4.0[e]	0.1 Very minor		0.2	18.0	0.1	0.05 Very minor	38.9
2. Public health									
a. Disease reduction (no./yr)		Major[b]	Very minor		Minor	100,000 (major)	?	Minor	Major[a]
b. Injury reduction (no./yr)	100,000	85,000 25,000[e]	0		Minor	580,000[a]	Minor	75,000	Major[a]
3. Safety									
a. Lives saved (no./yr)	?	5,000 minor[e]	Minor		Minor	39,000[a]	Minor	5,000	Minor[a]

b. Reduction of physical threat from criminal behavior	Low	Low to[c] mod.	0	0	0[c]	0	0	Low
4. Job opportunity	Very high	High	0	0	High	0	0	High
5. Attractiveness	Very high	High	Low	Low	High	Mod.	Low	High
6. Esthetic quality	Negative	Mod. to high	Moderate	High	High	High	Mod.	Mod.
7. Environmental quality								
a. Air quality (millions of tons—annual reduction)								
(1) carbon monoxide	1.6	1.6[f]	0	1.0	52.0	0	1.0	57.0
(2) sulfur oxide	2.5	2.5[f]	0	1.0	1.0	0	10.0	21.5
(3) nitrogen oxides	1.0	1.0[f]	0	1.0	5.0	0	2.5	8.5
(4) hydrocarbons	1.0	1.0[f]	0	1.0	10.0	0	1.0	15.0
(5) particulates	1.0	1.0[f]	0	1.0	1.0	0	2.5	9.0
b. Water quality	High	High	High	High	High	0	Depends	High
c. Relief from surface atmospheric conditions	High	High	Moderate	Mod.	High	Mod.	Mod.	High
8. Shelter								
a. Quality	Mod. to negative	0	0	0	Low	0	Low	Low
b. Availability	High	High	0	0	High	0	0	High
c. Density reduction	High	High	0	0	High	0	0	High
9. Mental health improvements	May be negative	?	0	0	Low	0	0	?
10. Ethnic quality	High	High	0	0	Mod.	0	0	Mod.
11. Conservation of natural resources	High	Mod.	Very high	Mod.	High	0	High[d]	High

[a] Depends heavily on taking advantage of technology for monitoring and control in the underground environs. [b] Depends upon technological use of underground advantages for sanitation and insect control. [c] These may be negative due to problems from combined vertical and horizontal patrol and control. [d] Depends upon elimination of coal and oil based generation, and no high stack needs. [e] Subdivided as follows: Production Systems and Structures; and Retail and Other Commercial Systems. [f] Effects of "Retail and Other Commercial Systems" included with "Shelter." [g] Except for transportation. *Source. Ibid.*, pp. 10-11.

of the savings. Time costs, largely from transportation, are the second largest source of savings.

As to life-support systems, shelter and production and industrial structures are the source of almost 50 per cent of the savings. Transportation is the single biggest beneficiary insofar as underground location is concerned.

Table 5.3 provides a good perspective of the potential economic savings from greater use of the underground. Furthermore, it also provides insight on the source of greatest savings, both in terms of type of cost and the life-support system.

Cost estimates of this type would not be sufficient as a basis for national policy—unless, of course, there was no better estimate. Preferably, more detailed evaluations should be developed as to the feasibility of transferring a life-support function to underground space. Furthermore, the technical feasibility of obtaining near-equal construction costs needs to be evaluated, or a reduction on the estimate of economic benefits would need to be made to offset the difference in construction costs.

Social Benefits

Preparing estimates of the social effects of locating life-support functions in underground space offered far more difficulties to the UCRC authors than did the estimates of economic savings. First, there were no generally accepted classification systems or measurement scales upon which to make an analysis. Second, interactions produced by the transfer are certain to occur, and a complex set of new relations are certain to develop. Estimates that did not attempt to deal at least in part with these interactions would have been meaningless. Accordingly, the UCRC study contained some judgments of the changes in relations that would develop.

SOCIAL INDICATORS

In the absence of generally accepted criteria, a set of social consequences were identified that appeared to be responsive to the aforementioned list of social problems. These indicators were identified as "measures of a state, condition, or movement of some factor." They are not to be confused with the objective of a social system; that is, like a thermometer that measures the temperature but does not tell what the temperature *should* be, the indicators cannot tell what the conditions *should* be. The indicators used in the UCRC study, many of which could not be treated on a quantitative basis, are summarized in the following paragraphs.

Opportunity time was defined as the man-hours saved. As one of man's most essential possessions, the time that is saved for an individual's pursuit of his own choosing is highly important. For the estimates prepared, a distinction was made between business travel, which had economic consequences, and pleasure travel. Typical time savings that were included resulted from (1) injury reduction, (2) relief from adverse environmental effects, (3) reduction of time for accident investigation, repairs, and such, and (4) increased reliability.

Public health was disaggregated and included savings from reduction of disease and of injuries. Although injuries are frequently considered under the label of public safety, in this study the latter label was restricted to loss of life and to threats from crime.

Safety benefits were defined as the reduction in loss of life and reductions in threats to "life and limb." Property losses were not included in this category of social consequences because they were estimated in economic terms.

Job opportunity was considered because the release of surface price (and space) constraints provides flexibilities and opportunities for job-creating establishments. Measures were necessarily crude, consisting of high, medium, or low effects.

Attractiveness, or the drawing capacity of an area, can be measured by population "ingress and egress" over time. This value was considered particularly typical of the inability of an indicator to be an objective. Thus to a Chamber of Commerce, population growth is good, but to many environmentalists, the same growth would be bad. Descriptive criteria, rather than quantitative measures, were used in the UCRC study.

Esthetic quality, like many other human factors, cannot be expressed in absolute terms. In the study, past and existing classification of the nature of "obnoxious" features were taken into consideration, because subsurface systems removed such facilities from the surface.

Environmental quality included air and water pollution and relief from surface atmospheric conditions such as heat. The removal of impurities was the primary measure of quality of air and water. Relief from adverse climatic effects, as offset by the absence of beneficial climatic effects, was considered, rather than esthetic or public health factors.

Shelter quality, both in terms of availability and density, was necessarily measured in descriptive impacts; that is, high, medium, or low.

Mental health was mentioned but evaluated very superficially. Further consideration is included in a subsequent discussion of adverse effects.

Ethnic quality indicators were necessarily gross and crude. Usually, employment coefficients, housing inequities, and such are used as indices, but these also are subject to complex limitations and interactions. Only descriptive criteria were used in the UCRC study.

Conservation of natural resources, as a social and not an economic factor, is typified by man's long-term interest in the value of water as a resource in addition to and aside from its economic value. The UCRC study considered only one natural resource, water, but suggested that there would be parallel findings for other resources.

ESTIMATES OF SOCIAL BENEFITS

The UCRC authors prepared estimates of the effect of moving 80 per cent of the life-support systems in the United States underground. Their detailed summary, contained in Table 5.3, is described as conservative but "full of problems." The major advantage to these data lies in the focus that they provide on ultimate benefits to individuals.

Certain indicators suggested major benefits. Thus of the 1.9 trillion man-hours that were available annually to Americans in the early 1970's—in which to do everything from sleeping to eating to playing—a minimum savings of 18 billion man-hours was estimated. Up to 50 to 60 billion could result, due to interactions. Although this may not seem large, it amounts to 150 hours for each individual per year or approximately half an hour per day.

Adverse Effects

The UCRC analyses of economic and social benefits were based upon incremental changes; that is, existing conditions were a reference, and the consequences were based on changes from that datum. Adverse effects, of course, were present for existing conditions, and the benefits that the report estimated were a reduction of existing adverse effects.

Making estimates of social consequences was more troublesome than for economic effects. Whether any change is an improvement or a degradation depends upon individual preferences.

Some adverse consequences would be introduced by underground construction. In the economic analysis, these took the form of (1) sunk capital in existing facilities, (2) economic dislocation, or change in who pays or who benefits, and (3) the allocation of resources that would be required for any major change in underground construction technology.

The adverse social effects have strong cultural overtones, relating to conditions in a society at a given time. For example, subsurface living would be considered by many individuals as a lowering of their basic life quality. This psychological effect is real and is treated accordingly; but it is nonetheless a transient characteristic that can change or be influenced.

Other features would be influenced by how rapidly a transfer was to be made, and/or the characteristics and use of subsurface space. Typical of facilities that would not cause too much difficulty would be retail stores, warehouses, manufacturing plants, office buildings, museums, and theaters. For these facilities, the interior conditions to which everyone is accustomed in present structures above ground would not necessarily be greatly changed for subsurface locations.

The most uncertain new consequence is the effect on human beings of prolonged periods of time in subsurface space. Little or no research has been done on the possible adverse or beneficial effects of prolonged stays in underground interiors that are comparable to or better than those on the surface. How can anyone, therefore, predict what effects long stays underground might have, until such stays underground can be studied and compared to stays in an identical surface interior? The full psychological and physiological effects of living underground are simply unknown. Not only that, the effect may be significant. (Incidentally, the same kind of uncertainty existed over tall buildings before there was experience with them.)

"Fear of the unknown" or simply a reaction against change is a most difficult adverse effect with which to contend. This factor is not unique to the question of surface versus subsurface space—any radical change needs to consider this factor.

Problems of second order effects became of more consequence in the midtwentieth century. Thus technology assessment and environmental impact analyses were designed to analyze the effect of new and existing technologies. The value derived from the results of such analyses is contingent on the ability to deal with interactions produced by a change, or series of changes. Predicting attitudes toward life qualities is difficult on a year-to-year basis. Such predictions for 5, 10, or 20 years in the future call for far more control of individual choices and actions than was possible in the 1970s.

With secondary effects it is important to (1) know of their existence, (2) monitor their effect, and (3) be prepared to take action promptly if necessary. To fail to reach for certain benefits because of uncertain or unknown dangers leads to a stagnation and sterility of life qualities. On the other hand, to reach for uncertain benefits in spite of reasonably predictable hazards is foolhardy. In either case, objective appraisals are essential to maximize the probabilities of improving the welfare of man, no matter how that welfare is defined.

EVALUATION OF ALTERNATIVE TECHNOLOGIES

The estimates of social and economic consequences provide insight on the question of whether greater use should be made of the underground. On the other hand, similar benefits are possible by making other changes in life-support functions. The UCRC study made only limited analyses of these other alternatives, anticipating that other studies of how to improve life-support functions would be available to decision-makers. To give a minimum perspective, however, the study presented the following analysis, given here only in brief summary.

Characteristics of Life-Support Functions

Most technologies for life-support functions, particularly the facilities, are based on old concepts or execution and are provided on the basis of the supply and distribution of resources relative to the consumer. The facilities are essentially a product of a balance between forces of technology, economics, and environment.

Until the early 1950s, technology played a major role in the decision of whether something should be built, and particularly how it should be built. As the total consequences became of greater social interest, the surface construction of life-support functions was observed much more carefully by the general public. Natural resource depletion, energy requirements, pollution control, safety, public health, and social disruption became as important as, or more important than, economic growth. The consideration of greater use of underground space is thus a question of whether the adverse effects of surface construction can be offset by a subsurface location. Obviously, there are means other than going underground to reduce the adverse effects, and all valid alternatives should be considered.

To compare alternatives in a rational manner, criteria are needed for which the measures are commensurate. Once these are set, the dollars required to produce a unit of change in the criteria can be estimated. This is, in effect, a cost effectiveness approach.

A second technique was employed in the UCRC study because of the absence of data and the desire to provide to the decision-maker other bases on which to consider the alternatives. The procedure, termed priority effectiveness, was based on a subjective appraisal of technical, social, economic, and political feasibility. The two techniques, described in the pages that follow, were based on a subjective view of alternatives, as shown in Table 5.4.

Cost Effectiveness Analysis

This evaluation consisted of two steps: (1) for each function, each alternative was evaluated against 19 criteria (Table 5.3); and (2) for each al-

Table 5.4 Alternative Procedures for Improving Civil Works Functions Both Underground and Above Ground

1. Liquid waste disposal and treatment alternatives
   a. Improved large scale treatment facilities
   b. Population distribution
   c. Distributed treatment
   d. Improved underground construction
2. Solid waste management
   a. Improved large scale conversion
   b. On-site compaction, reducing collection costs
   c. On-site conversion, e.g., wet oxidation
   d. Underground collection system, e.g., Centrasug
   e. Rapid biodegradable packaging materials
3. Energy distribution
   a. Unit generation
   b. Reduced demand
   c. Underground distribution system
4. Communication
   a. Underground cable
   b. Satellite transmitters and repeaters
   c. Improved transmission controls
5. Water resources
   a. Unit closed cycle systems—no distribution
   b. Improved underground distribution technology
   c. Redistribution of population to balance water supply and demand
   d. Large scale diversion of supply to meet demand
6. Shelter, commercial, industrial, and production structures
   a. New materials for lower cost construction
   b. New methods of construction to lower time and cost of construction
   c. Concentration of people in multiple dwelling units rather than individual ones
   d. Locating dwelling and other "housed" functions underground
7. Transportation
   a. Mass transit
   b. Substitution of communication for physical movement
   c. Alternative forms of propulsion and energy supply
   d. Locate existing technology in subsurface space

Source. Ibid., p. 16. (see also Table 5.5).

ternative that showed a significant improvement as measured by the criteria, the requirements were evaluated.

A detailed analysis of each of the alternatives listed in Table 5.4 would have been better. The UCRC study, however, was not able to include more than a global, descriptive, judgmental approach. This is a reasonable tactic in policy analysis wherein allowable time and resources determine the depth of study. The following is excerpted from that report to illustrate the descriptive character of the evaluation that is frequently necessary for policy analyses.

Liquid Waste Disposal Treatment

A large national investment is now being made in waste water collection and treatment. This is largely focused on (1) preventing untreated waste disposal by legal sanctions, and (2) funding for the installation of sewer systems and treatment facilities. These will improve environmental quality, but only with a significant long-term increase in public investment. The largest element of that cost is in *collector sewer systems*. A more efficient sewer construction system would reduce both cost and time to attain, nationwide, a socially desirable water quality level. In fact, those levels cannot be achieved unless and until the collector sewer systems are installed. Hence, *improved waste water treatment facilities—although necessary—are not the determining factor in attaining water quality goals*. Population redistribution would, of course, reduce the demand on the environment, both natural, and man-made. It would, therefore, diminish water pollution rather quickly.

Instead of using sewer systems to transport the wastes to a treatment facility, each waste producing unit could treat its own. There have been three different types of systems developed and proposed for accomplishing this: wet oxidation, reverse osmosis, and gamma ray irradiation. Real promise exists for these methods since they can be used not only in new construction, but also by retrofitting. Their use depends, in part, on economic feasibility, but they have the potential for providing a far higher level of water quality control at an equal or lower unit cost than existing technology.

Improved underground excavation can, in conjunction with planned waste water treatment facilities, increase the speed and reduce the costs of building sewer lines. This would lead to attainment of higher water quality standards sooner and at significantly lower costs.

Of the four alternatives, the order of priority with which each would produce the greatest benefit in terms of the nineteen criteria are: (1) distributed treatment; (2) improved underground construction; (3) population redistribution; and (4) improved treatment facilities.

Solid Waste Management

Vast improvements have been and are being made in the large conversion devices currently in use. Incineration has been more highly developed to meet air pollution standards and costs have soared with these improvements. Land fill (sanitary) operations have been improved and completed land fills are being used as recreation sites. Cost savings or benefits from these measures are primarily social, and increased costs using these conversion methods will be unavoidable if federal standards are to be met.

On-site compaction is available for both residential and industrial use and can reduce the frequency of collection. The question is, how will this affect collection costs—the major costs associated with solid waste disposal? Most collection vehicles today are equipped with compactors so that on-site compaction would primarily shift the time and cost of that operation to the individual solid waste producer. This would reduce the capital investment in collection equipment (elimination of compaction equipment) and reduce the labor involved because less bulk would have to be handled per producing unit.

On-site conversion could take several forms. The garbage disposal (grinder) common in many households today merely converts a portion of the solid waste into a form compatible with the liquid waste disposal system, thereby transferring the responsibility and cost for disposal to that system. Any process such as this could reduce or eliminate the collection phase of solid waste disposal with a consequent reduction in cost. The trade-off is between the total cost of the two systems and the fact that the cost of on-site conversion units can be shifted from the public to the private sector.

An underground collection system should provide considerable operating cost reductions. Capital costs would be reduced by development of better underground construction technology and the labor intensive and high capital maintenance cost of present surface collection systems would be eliminated. Improved sanitary conditions and general cleanliness should also result from elimination of present multi-handling collection systems.

Rapid biodegradable packaging materials would reduce incineration and landfill problems, and should reduce the size and complexity of conversion systems. The costs for such materials may, in many instances, be quite high. Consequently, it is difficult to estimate the cost effectiveness of the approach.

Evaluating each of the alternatives and recognizing the interdependence among several of them, the pay-off would appear to be: (1) on-site conversion, (2) underground collection systems, (3) on-site compaction, (4) improved large-scale conversion, and (5) rapid biodegradable packaging material.

Energy Distribution

Unit generation eliminates the distribution of electrical energy by replacing large generating units with small, on-site generators. The economics of this will depend largely on the power source for unit generation. If this source is fuel oil or gas, the electrical distribution system is eliminated only to be replaced by the fuel oil or gas distribution system which has similar distribution problems. Some sort of energy distribution seems inevitable and benefits will arise from limiting the distribution process to as few materials (energy sources) as possible, based on an optimizing analysis of the merits of the generating and distribution characteristics on each energy source.

Reduced demand could result from a shift of energy sources such as a trend away from the "all-electric living" concept or a change in living patterns. The first would merely shift the source and the arguments concerning unit generation would apply. The second would be accomplished by eliminating or reducing climate control systems or rearranging activities and designs to make maximum use of daylight hours, or restricting development of geographical areas requiring high degrees of climate control. Social patterns would have to be overcome and "progress" might have to be curtailed somewhat.

Undergrounding of the distribution system would both eliminate esthetic conflicts with the surface environment and decrease operating costs. The priority for these alternatives is (1) undergrounding, (2) reduced demand, and (3) unit generation.

Communication

By undergrounding communication cables, maintenance requirements and service interruptions due to environmental interference are reduced. Likewise, conflict with other surface space uses and questions of esthetics are eliminated. The sharing of undergrounding costs through the use of a common utilities tunnel would likewise be applicable to this.

The use of satellites in communications is now a reality, possibly the best known use being Telestar for overseas transfer of television transmissions. For long-range communications, benefits are derived by the elimination of physical right-of-way (in the sense of land and physical materials). With the exception of the satellites, all equipment subject to breakdown is located in centralized, accessible places. For short-range use, the equipment costs would have to be compatible with surface distribution costs. Technology transfer from the space program can greatly benefit this area.

Greater benefits will accrue as technology becomes available to increase transmission densities on existing equipment (cables, microwaves, etc.). Thus greater density of communication channels would be provided on existing distribution systems, thereby reducing the total distribution space—either spectrum or cable—required.

The order of priority would be (1) satellite transmitters and repeaters, (2) improved transmission controls, and (3) undergrounding cable.

Water Resources

If a unit closed cycle system can be devised that will have negligible loss, distribution would indeed be eliminated and great benefits would accrue. However, losses will occur in the system so that a distribution system for recharging will be required. Obviously, the total costs will depend on system efficiency and recharging rate.

Undergrounding improvements will render benefits both in lower capital costs for system extensions and in reduced maintenance and operating costs. Joint use of a transportation and/or joint utility tube would provide both accessibility and further capital savings.

Redistribution of population would provide considerable savings in distribution costs, but must be balanced against social resistance and possible increase in other costs. Industry often seeks out new locations on the basis of available water and further control of this could encourage more population redistribution or at least limit expansion of areas in which water supply is a problem.

Large scale diversion of supply is costly, though the practice is used. Improved underground technology would produce savings for water distribution systems since tunneling is currently required. Furthermore, undergrounding reduces surface losses (evaporation) and pollution during long-run diversions.

Priorities would be (1) improved underground distribution technology, (2) unit closed cycle systems, (3) redistribution of population, and (4) large scale diversion of supply to meet demand.

Shelter, Production, and Commercial Structures

New materials which would lower cost would have a beneficial effect on the economy as the products would have to be developed and new plants built for their production. It is possible that these new materials might be produced from some industrial by-product, thus reducing a waste disposal problem and conserving the natural resource which it replaces.

By developing new construction methods, or by applying tested industrial procedures to the construction industry, significant cost and time reductions could be realized, thus freeing resources for other needs.

Concentration of people in multi-dwelling units would produce savings in land and all services to the buildings (electricity, sewer, water, waste disposal) and for certain size structures, would result in a reduction in total building cost.

The location of structures underground results in a multitude of savings, the most influential being land and energy requirements. Here, it is the sociological effect that must be overcome, for open space and at least visual exposure to the atmospheric environment is an important attribute to most individuals.

The order of priority would be (1) new methods of construction, (2) new materials for lower cost construction, (3) concentration of people in multiple dwelling units, and (4) undergrounding of structures.

Transportation

The benefits of mass transit lie primarily in the reduction of right-of-way costs due to the denser utilization of space. Other factors to be considered include conservation of fuel resources and reduction of pollution resulting from consumption of these resources. However, the costs of these systems for extended metropolitan areas is very high and provision of all urban transport services will require a number of vehicles approaching some limit close to that of the number of automobiles.

Substitution of communication for physical movement completely eliminates the need for transportation corridors and equipment for the purpose of interpersonal communication and exchange of ideas. However, a transportation system for the movement of physical goods would still be required. Furthermore, the roadway capacity, and the switching and display sub-systems that are required are very large and cost commensurately greater.

Alternative forms of propulsion and energy could result in more efficient and less costly transportation equipment, possibly making new forms of transportation feasible. The physical facility system will still be required.

By undergrounding highway transportation, the numerous conflicts with adjacent land space could be eliminated, and much land left available for other uses. Major savings are in travel time, reduced pollution costs, and reduced disruption of surface space.

The priority order among these alternatives is (1) undergrounding of the transportation system, (2) substitution of communication for physical movement, (3) mass transit, and (4) alternative forms of propulsion and energy supply.[2]

All of the alternatives listed in Table 5.4 were considered to be possible, and a method of comparison was established. The four criteria that were selected were (1) technological feasibility, (2) time to implement, (3) resources required to implement, and (4) social and political feasibility.

A judgmental technique was used as shown in Figure 5.2. This approach provided a metric that was commensurate across each dimension. Hence it provided a means of evaluating each alternative and combining ratings.

This is a useful technique for policy analyses, but must be utilized with care. Within each of the four dimensions, there is no abstract or idealistic measurement problem. To sum the four dimensions, however, requires one

[2] *Ibid.*, pp. 15–21. Italics in original.

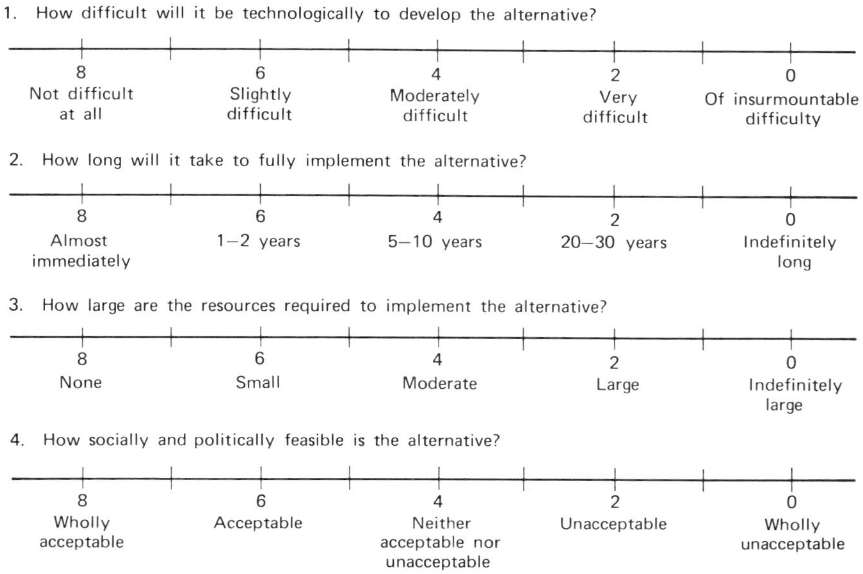

Figure 5.2. Criterion scales for feasibility measures of alternative procedures. *Source. UCRC Report*, p. 22.

of two assumptions: (1) that they have equal weight, or (2) that an acceptable weighting system can be developed.

The UCRC rating for overall cost effectiveness produced the results shown in Table 5.5, wherein equal weight for the four dimensions was assumed. If we combine the results of Table 5.5 with those from Table 5.3, transportation emerges as the function which combines the highest economic return with the highest feasibility. Undergrounding electrical energy distribution would provide the next highest economic return and feasibility. The analyses are incomplete as to the economic return for alternates other than underground construction. But for the underground alternative, a ranking is available.

By assigning a higher weight to any of the four dimensions (Figure 5.2), a decision-maker can obtain a different perspective. If desired, the sensitivity of the results to the weighting could be developed.

Priority Effectiveness Analysis

In the second technique, an evaluation was made of the relative importance of transferring various life-support systems to the underground. The

Table 5.5 Rating of Alternative Procedures of Improving Civil Works Function Performance (High Numbers Being High Ratings)

	Alternatives				
Civil Works Functions	a	b	c	d	e
Liquid waste	18	7	18	**18**	—
Solid waste	15	18	19	**15**	16
Energy distribution	11	13	**18**	—	—
Communication	**20**	26	18	—	—
Water resources	12	**19**	8	11	—
Shelter, commercial, industrial, and production structures	23	19	12	**7**	—
Transportation	10	16	15	**18**	—

Source. Ibid., p. 24.
Note. See Table 5-4 for detailed description of alternatives, matched to letters a through e. The rating for the underground alternative is in bold type in this table.

following procedure was used:

- Those functions were selected that could, with modification, provide advancement toward social, economic, and cultural goals.
- Those functions were selected that could, by use of the present state of technology or by modification and upgrading of technology, be located and maintained in whole or in part in the underground.
- Those functions were selected that would have application nationally in terms of magnitude, complexity, and cost as well as functional necessity.
- The functions were described, providing some of their most salient characteristics.
- The gains and losses were estimated that would accrue from locations in the subsurface rather than on the surface.
- The definition of feasibility was expanded to include measures (or indicators) of social, environmental, and political feasibility as well as the traditional economic and technical indicators of feasibility.
- The developed representation of feasibility measures was applied.
- A method for a comparative evaluation of overall feasibility was developed.

- The results were interpreted and compared with results from the cost-effectiveness analysis.

Three judges produced a subjective rating based on the data in Table 5.3. To obtain commensurate values, rankings were made on a scale ranging from 1 to 10 for the life-support functions in terms of social, environmental, economic, and political feasibility. Table 5.6 provides a summary of the results.

Four observations can be made relative to these results:

- All functions that have high social, economic, and environmental gains have low technical and/or political feasibility, except for transportation.
- Transportation is a prime function for locating in subsurface space in the near future.

Table 5.6 Ratings for Priority-Effectiveness Analysis

	Feasibility Types			
Civil Works Functions of Life Support	Social	Environmental	Economic	Political
Shelter	7	7	7	1
Commercial	5	7	6	4
Production systems	5	7	6	2
Water resources	1	5	2	9
Liquid waste	1	6	2	9
Solid waste	2	5	3	4
Transportation	7	8	9	7
Communications	3	1	2	8
Energy distribution	2	7	3	7
Storage	2	2	2	8

Source. Ibid., p. 26.
Legend. 1—Low feasibility, 10—High feasibility.

- Some of the most technical and politically feasible alternatives do not involve the largest gains, but the gains are still appreciable.
- The largest gains to the environment and to man himself accrue from the potential transfer of (1) transportation, (2) production systems and industrial structures, (3) human shelter and housing, and (4) storage. The gains are in dollars, lives and time saved, and in the reduction of injuries, disease, and pollution.

The ranking that produces results of the type used in this technique necessarily requires (1) weighting of relative importance of different types of feasibility or (2) subjective weighting by the decision-maker. Thus a high (or low) score on one type of feasibility (e.g., political) may dominate one decision-maker's view while for another, a different feasibility (e.g., environmental) might be dominant.

TECHNICAL CAPABILITIES

The preceding discussion has been directed to whether life-support functions *should* be transferred to the underground. Prior to setting specific goals, the question of whether such a transfer *can* be made needs to be considered; that is, whether construction technology is adequate.

State of the Art

The UCRC study included a comprehensive evaluation of the state of the art. This included an analysis of the costs, rates of advance, and other characteristics related to the health and safety of the workers.

The costs of tunneling are quite variable, and any average unit cost is practically meaningless. Figure 5.3 illustrates the variations. To provide some perspective on unit costs, the report contained an analysis of 112 tunnels completed (or near completion) in the period 1960 to 1972. The total length of these tunnels was 869,091 feet and the total construction cost was $720.1 million. The average bid price was $828 per lineal foot ($4.4 million per mile). For the tunnels surveyed, the following approximate costs were indicated:

Excavation and temporary support	$260 per lineal foot (27 per cent)
Permanent lining	$140 per lineal foot (15 per cent)
Other installations	$570 per lineal foot (58 per cent)
Total bid price	$970 per lineal foot (100 per cent) or $5 million per mile

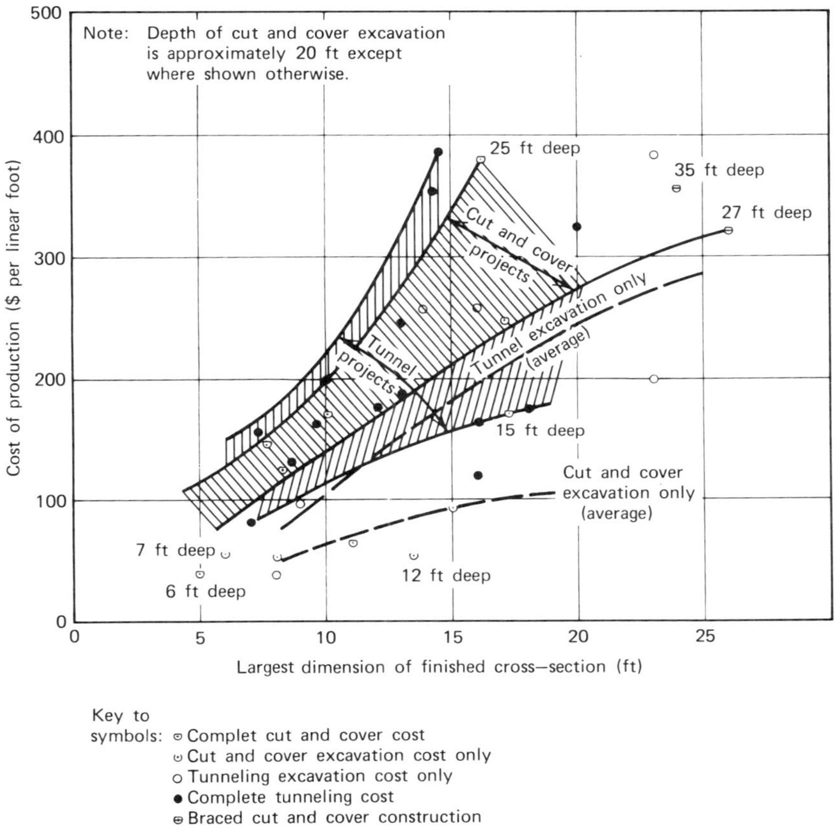

Figure 5.3. Comparison of underground construction methods—cut and cover versus tunneling. Data developed from returns of UCRC questionnaire, Fall 1971. *Source. UCRC Report*, p. 31.

The rate of advance is also quite variable and is a key factor in cost. Figure 5.4 shows the wide range in rate of advance, measured in feet per 24-hour day. A value of 30 to 40 feet per day covers a reasonably wide range of conditions.

The analysis of the state of the art produced a series of conclusions. *First*, significant improvements had been made in the technology during the late 1950s and 1960s. The major advance had been made in the development of tunnel boring machines which permitted full-face excavation at steadily increasing rates of advance. The machines were most effective for tunnels less than 20 feet in diameter.

Second, the underground excavation process was a multistage one, rather

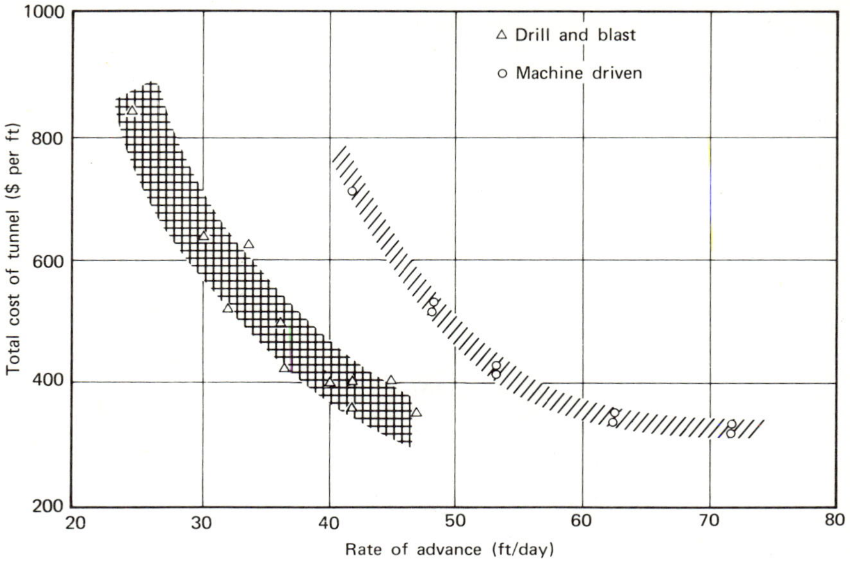

Figure 5.4. Comparison of rate of advance and total cost (example tunnels in Rock). *Source. UCRC Report,* p. 32. From "Some design considerations in the selection of underground support systems," University of Illinois for office of High Speed Ground Transportation, USDOT, Contract 3-0152, November 1969.

than a continuous operation; that is, it consisted of a cycle of (1) excavation or boring, (2) muck removal, and (3) tunnel lining.

Third, the needed improvement in rates of advance and costs was quite variable, and the most effective changes that could be made would require a detailed technical evaluation.

A general overview of the state of the art was that the existing underground construction process did not lend itself to a major, accelerated transfer of life-support functions to the underground. To achieve such a transfer within the foreseeable future, a new technology would be required—a technology that would close the gap between the costs and time for surface and subsurface construction.

Inventory of Research and Development

Industry and government research and development were inventoried. Relatively complete government data were compiled from standard sources. However, it was necessary to make a survey to obtain data about industry. Queries to 670 private organizations produced 293 returns, on the basis of which industry activity was estimated.

The inventory indicated that funding for research and development was increasing in both the public and private sectors. Federal government programs were expected to nearly double in 2 years from $19 million in 1971 to $36 million in 1973, if the President's budget was funded. Private sector estimates were at the $7 million annual level in this same period.

The subject matter of research and development varied widely in the federal government. The program was fragmented, being a function of the various agency priorities.

GOALS

The preceding analyses provided a series of alternative program requirements, and produced a goal that was based on the assumptions of the study and on the following:

- Substantive benefits could accrue nationally by greater use of underground space.
- To obtain the benefits, a policy decision would be needed that these benefits and the method of achieving them had sufficient priority and were in the national interest.

The UCRC study proposed that the federal government adopt a general policy to accelerate the transfer of life-support functions to the underground. The sequence and timetable for various functions would be determined after the policy was adopted. One of the alternative requirements for achieving the transfer was an accelerated research and development program to provide a new technology. This requirement was stated as follows.

> Develop an underground construction process that approaches equivalency with surface construction in terms of (1) cost, (2) time to construct, (3) safety during construction, (4) working conditions, and (5) indirect effects of construction.
>
> The range of funding level to fulfill the preceding requirement is from 0.1 to 1.0 billion dollars per year for five to ten years.

The preceding goal was adopted by the Underground Construction Research Council. Time-cost analyses were also conducted (see Chapter 3) to provide additional data as to feasibility. The UCRC then proposed that specific national goals be established for transferring life-support functions to the underground.

The preceding requirement for research and development would thus become an integral, alternative part of the implementation of the transfer; that is, the national goal would be stated in terms of transferring life-support functions that would assure the benefits. The research and development would become a means to an end, rather than an end in itself.

Two national programs were analyzed in some depth: (1) urban expressways and (2) urban electrical energy distribution systems. Specific programs in these areas, developed as examples to demonstrate how national policies would be influenced, led up to the following conclusion.

> It is concluded in the report that firm commitments to use underground space to achieve major social, economic and environmental objectives will be required to stimulate greater use of the sub-surface. The most promising (and most feasible) of such programs is proposed as a short range goal (immediate future); namely, a requirement that underground locations be used for new urban expressways and urban electrical energy distribution systems. It was estimated that an annual $1.55 billion program for ten years could provide, in underground space, 5000 miles of urban freeways and two-thirds of a million miles of electrical energy distribution lines. By the end of ten years, the total $15.5 billion investment could be recovered (through benefits) and released for other uses, and a $6 billion annual economic savings would accrue thereafter. Additional social and environmental benefits would result from release of surface space, improved facilities, less disruption of neighborhoods, and an action program directly responsive to urgent concerns of urban areas.
>
> The annual $1.55 billion cost for the short range program includes $150 million for an integrated research, development, testing and evaluation program, and $50 million for managing the implementation. Specific research needs are dependent upon the requirements to achieve national goals and upon the level of resources and timetable. The results of an inventory of on-going R&D are presented and the problems with the highest technical priority are listed.
>
> What is now required is a policy decision that the attaining of these benefits has a sufficiently high priority.[3]

The study and the ASCE actions that followed (recommendations on a national goal) were unique in several ways. First, specific goals were identified, rather than nebulous guidelines. Second, the benefits of the proposed program could be readily analyzed by the political system, and compared with benefits from quite different types of activities. Third, maximum opportunity to utilize science and technology would be possible because the goals were sufficiently specific, but not overly restrictive.

[3] *Ibid.*, summary, p. iii.

CHAPTER 6

POLICY DEVELOPMENT FOR ESTABLISHING CONSTRAINTS—ENERGY CONSERVATION IN TRANSPORTATION

The previous examples in Chapters 4 and 5 were concerned with how to allocate economic resources to produce an optimum advance in the life of a community. The example in this chapter is typical of two other types of applications of policy development: one, the allocation of other than economic resources; and two, the use of constraints as a method for allocating resources.

As suggested in Chapters 2 and 3, a number of types of resources are available to a society, including, manpower, energy, land, and natural resources. Allocation of these resources has traditionally been handled in the United States through public utilities and the market place. In times of emergency, rationing has been imposed by the government.

When in the public interest the government concludes it must intercede in the allocation of other than economic resources, the same type of problem arises as has been previously discussed; that is, how can the particular resource be allocated so as to produce the optimum life qualities for the community.

This chapter also illustrates how constraints can be used to allocate resources, as opposed to providing resources to specific programs. As is evident, this is in principle, the same type of action in that resources are assigned to one activity through the withholding from another. Because a choice must be made between alternative benefits, this type of policy involves resource allocation for which policy development is needed.*

THE POLICY ISSUE

To proceed let us suppose that it is 1960 and careful analysis has shown that in the United States the probable supply of indigenous petroleum will not meet expected demand by 1973. Since the United States is energy dependent, it appears essential that both alternative means of supply and a reduction in consumption to meet available supplies is necessary by that time. Since about 30 per cent of all petroleum production is used for transportation purposes, it seems apparent that reduction in transportation consumption produces significant oil savings.

From a public policy standpoint, it is important to make the most efficient use of the energy supply to retain the integrity of the social and economic structure which is dependent on energy. In the face of a diminishing supply, it should become a public policy goal to conserve the supply, and/or make the most efficient use of that supply, and/or create al-

* Chapter 6 is based upon a report, *Energy Requirements in Urban Transportation* by R. M. Michaels and A. B. Maltz, Proc. 1st Ill. Energy Conf. Chicago: UICC, 1974; 247–270.

ternative sources of energy to meet projected social needs. The problem then becomes one of determining the best way to achieve this goal.

Now ideally, it would be nice to be able to define the minium reduction in consumption desired from any sector of transportation. However, there is no way for the policy-maker to know what the effects of any reduction may be, direct and indirect. Nor does he know even how much a reduction in consumption is possible. Finally, it is obvious that the predicted deficit suggests a target that should be met. Hence at this stage of the policy development process, it is necessary to provide only a rather general goal statement. The policy issue now becomes one of defining the alternative means, social or technical, that will produce a significant reduction in gasoline consumption in transportation.

This policy issue is, of course, only part of a larger policy goal, that is, to ensure an adequate and economic supply of usable energy. A complete and comprehensive policy development process would be concerned with all elements of the chain from resources to end uses. However, from a public policy standpoint, the uses of energy involve different sets of societal systems, most of which have very different requirements than do those which provide the raw materials and transform them into usable forms of energy. Energy production and distribution is an infrastructure system. It is important because of the way the social system has organized on the assumption of its continuing plenitude. If, or as, it becomes scarce, it threatens the integrity of the society. From a public policy standpoint, it is that integrity which must be protected, preferably before the scarcity becomes so great that social and economic disorder becomes unavoidable.

To do this the public policy-maker must know what the supply situation will become and what effects any shortage will have on the different societal systems. In addition, he must develop control policies based on social requirements and goals. It is these goals that determine the forms of energy resources that will be required *and* the structural systems of use required to ensure efficient use and continuing availability. Conservation of energy is, then, one important avenue to achieve this end.

Conservation is applicable to four major classes of societal technology: transportation, shelter systems, industrial production systems, and energy production systems. Each incur major losses, that is, have basic inefficiencies, which may be reducible. Transportation was selected for this analysis simply because it is an important user of energy and it is a single infrastructure service system that can be treated as an independent societal system. In the larger public policy development process, each of the other energy consuming systems would be evaluated in the same fashion. All would then be combined into an overall energy conservation policy decision. The policy development process would include the evaluation of alternative strategies for production and distribution. Thus the complete policy development process would develop an overall national energy policy. It would be a program that was time phased and cost estimates prepared so as to ensure that the necessary minimum quantities of energy would be available consistent with decreases in supply of current forms of energy. The end result or purpose of this policy would be to constrain activity so as to ensure the presence of the resources needed to maintain the integrity and natural growth of the society toward its human goals.

POLICY ANALYSIS

In the context of this policy issue the analysis is concentrated on a single societal system. Basically, it is concerned with a series of alternative means of manipulating that system. It seems reasonable first to identify the uses of energy in the transportation sector (shown in Table 6.1). As may be seen, 90 per cent of all transportation energy, mainly petroleum, is expended in five domains: (1) urban auto, 31 per cent; (2) intercity auto, 26

Table 6.1 Percentage Transport Energy by Use

Use	Percentage of all U.S. Transport Energy
Intercity, rural auto	26.4
Air	11.4
Intercity rail	3.5
Waterway, pipeline	1.4
Intercity bus	0.2
Intercity truck	7.0
Local truck	14.9
Urban auto	30.7
Urban rail	0.1
Local bus	0.6
Other	3.8
TOTAL	100.0

per cent; (3) urban truck, 15 per cent; (4) intercity air, 11 per cent; and (5) intercity truck, 7 per cent. Thus two-thirds of all energy expenditure in transportation is for person movements by automobile or air. Most of the remainder is for goods transport.

Projected estimates using a 4 to 6 per cent growth in demand indicate a requirement in the United States for approximately 15 million barrels of petroleum a day by 1970 and 18.5 by 1975. Furthermore, by 1970 between 20 to 25 per cent will have to be obtained from foreign sources while with projected growth rates, 40 to 50 per cent will have to be imported by 1975. Given worldwide growth in demand and potential disruptions in the supply chain, it is also projected that shortfalls in supply can reach 10 to 20 per cent in the early 1970s. Since transportation is almost wholly dependent on petroleum, it is clear that reduction in demand for petroleum would reduce or eliminate the effects of such a deficit. From a practical standpoint, the major source of demand elasticity lies within the area of person movements. Approximately 20 per cent of all petroleum use is for this purpose. That is, around 3 to 4 million barrels of oil a day will be expended for person movements in the 1970 to 1975 time period. Policies that reduce this expenditure can, over the short run, reduce markedly the likelihood of a social and economic disruption due to a shortfall in this period. It is also clear, that for the long run, beyond 1975 to 1980, alternative sources of energy must be developed and implemented to minimize the probability of major physical, economic, and social disruption.

IDENTIFICATION OF ALTERNATIVES

Since the policy issue deals with a single societal system, it is clear that the analysis is centered on those policy dimensions which may decrease utilization of petroleum within that system. Furthermore, the area of transportation where conservation is most possible is in person movements. A first step, in policy analysis then, is to examine each of the means of modifying the three major person movement elements of the transport system. It is then necessary to identify alternative means of reducing usage or increasing efficiency of use. In Tables 6.2 through 6.4 for urban and rural auto and air transport, a series of conservation strategies is identified. The list of alternatives is reasonably comprehensive, although others can be added. It is also worthwhile to note that there is considerable redundancy across the three transport systems. This is to be expected and, in fact, indicates that certain conservation alternatives can be applied over a large segment of the person movement sector. It may further be noted that several of the alternatives interact. Economic policies are aimed less at

Table 6.2 Alternative Means of Reducing Petroleum Use in Urban Automobile Transportation through Various Modifications

Institutional Rearrangements	Legal Sanctions	Economic Rewards	Value Changes	Technological
1. Land use	1. Speed limits	1. Gas taxes	1. Change of preferences from auto to mass transit	1. New power source
2. Regional mass transit	2. Rationing	2. Horsepower taxes	2. Car pooling to reduce travel	2. PRT (personal rapid transit)
			3. Curtail travel	3. Traffic flow control
				4. Substitution of communication for travel
				5. Smaller vehicles

Table 6.3 Alternative Means of Reducing Petroleum Use in Rural Automobile Transportation

Institutional Rearrangements	Legal Sanctions	Economic Rewards	Value Changes	Technological
1. Improved delivery services	1. Speed limits 2. Rationing	1. Gas taxes 2. Horsepower taxes	1. Reduction of travel	1. Power sources 2. Substitution of communication for travel 3. Smaller vehicles

Table 6.4 Alternative Means of Reducing Petroleum Use in Air Transportation

Institutional Rearrangements	Legal Sanctions	Economic Rewards	Value Changes	Technological
1. Change in service criteria 2. Carrier consolidation 3. Nationalization	1. Speed limits	1. Reduction of competition 2. Subsidization	1. Reduction of use	1. More efficient aircraft performance 2. More efficient ATC

revenue production than at reducing travel. Production and utilization of smaller vehicles would reduce the amount of gasoline used and incidentally would reduce the effects of any tax penalty policy.

We treat each of the three person movement domains independently. Essentially, they are functionally independent so that change in one has little effect on the others. One result of this assumption is that any petroleum savings obtained in one system are additive to savings obtained in another. The only exception may be any long distance trips generated in urban areas but with destinations in rural places.

ANALYSIS OF URBAN TRANSPORTATION ALTERNATIVES

Given the alternatives for conservation indicated in Tables 6.2 through 6.4, it is possible to identify their sensitivity to social criteria using the procedure from Chapter 3. As may be recalled, there are 12 such criteria. A rating scale approach is used where +3 denotes a large improvement in that criterion dimension and where −3 is an estimate of a large decrease in the criterion dimension. Using an appropriate panel of knowledgeable people, an estimate of the effectiveness of each of the alternatives is obtained, such as that given in Table 6.5, which shows the rating for the 14 possible ways to obtain energy conservation in urban transportation.

All the alternatives are perceived to generate energy savings or benefits, and there are significant indirect effects. Most of the indirect effects are positive, but there are also some interesting negative ones. The institutional, legal, economic, and value change alternatives reduce the efficiency of time utilization more than do the technological alternatives. That is, they require significant increases in travel time over the present system. Also, tax policies and rationing are believed to affect ethnic quality negatively, because they place a burden on those with already inadequate transportation. In essence, such policies lack equity.

Of the 14 alternatives evaluated in Table 6.5, the range of scale scores is from 0 to 9. Without proof we can conclude that the 4 top-rated policy alternatives are significantly more effective than the remaining 10. Hence, these will be included for detailed evaluation. Three others are in the process of being implemented: regional mass transit, car pooling, and smaller vehicles. Consequently, they too will be included for further analysis.

Economic Evaluation

If any of the seven selected alternatives were implemented, what would be the net benefit to the society? This is a second question that policy analysis

Table 6.5 Effects of Transport Energy Conservation Policy Alternatives in Urban Transportation[a]

	Institutional Rearrangements		Legal Sanctions		Economic Rewards		Value Changes			Technological Changes				
Social Criteria	Land Use Re-arrange.	Re-gional Mass Transit	Speed Limits	Ration-ing	Gas Taxes	Horse Power Taxes	Prefer-erence Changes to Transit	Car Pooling	Curtail Travel	New Power Sources	PRT	Traffic Flow Control	Substitu-tion of Com-munica-tion for Travel	Smaller Vehicle
Environmental														
1. Pollution	+1	+1	+1	+1	0	0	0	+1	+1	+2	+3	+1	+1	+1
2. Esthetic goal	0	0	0	0	0	0	0	0	0	0	0	0	0	0
3. Land use	+2	0	0	0	0	0	0	0	0	0	0	0	0	0
4. Biotic	0	0	0	0	0	0	+1	+1	0	0	0	0	+1	0
Resources														
1. Materials	0	+1	0	0	0	0	+1	+1	0	−1	−1	+1	+2	+1
2. Energy	+2	+2	+1	+2	+1	+1	+2	+2	+2	+3	+1	+2	+2	+3
3. Time	+2	−1	−1	−2	0	0	−1	−1	−2	0	+1	+1	+1	+1
Social														
1. Safety	+1	+1	+1	+1	0	−1	+1	+1	+1	0	+2	+2	+2	−1
2. Public health	+1	0	0	0	0	0	0	0	0	0	0	0	0	0
3. Shelter	−1	0	0	0	0	0	0	0	0	0	0	0	0	0
4. Mental health	−1	0	0	0	0	0	0	0	0	0	0	0	0	0
5. Ethnic quality	+1	+1	0	−1	−1	0	0	0	0	0	+1	0	0	0
Totals	8	5	2	1	0	0	4	5	2	4	7	7	9	5

[a] +3 = large improvement
0 = no change
−3 = large decrease

should answer. To do this, the procedure developed in Chapter 3 is used. In effect, we "monetize" the direct and indirect effects of each of the alternatives. It is appropriate to describe each alternative and its operational properties.

RELOCATION OF RESIDENCES RELATIVE TO WORK PLACES

Since 50 per cent of all urban trips are work trips and these trips account for approximately 75 per cent of the total vehicle-miles of intraurban travel, anything that reduces the median trip length must yield significant savings. One way in which this can be accomplished is to ensure that residential location choices are made in relation to work place location, or in other words that people live close to work. If it were possible to control residence location so that no worker was more than 5 miles from his work site, then the median work trip length would be reduced by one-half to two-thirds.

OPTIMIZE TRAFFIC FLOW

Some of the most significant wastes of energy occur as a result of traffic congestion. Energy consumption is highest under conditions where automobiles (and buses) must make repeated accelerations. The loss is even higher for vehicles idling under stopped traffic conditions. If the flow of traffic can be smoothed and stops minimized, significant savings in energy consumption may be obtained. These benefits are potentially available with the institution of better street capacity utilization, area-wide traffic control systems, and electronic aid systems in the highway and the vehicle. Work over the past decade suggests that such systems are within the state of the art and can be installed in existing highways and vehicles. Such systems as automatic routing systems, intervehicle spacing control systems, ramp metering and merge control systems, and network traffic surveillance and control systems are not cheap except relative to new highway construction.

INCREASED UTILIZATION OF MASS TRANSIT

From the standpoint of energy consumption, mass transit is attractive because it is a relatively insensitive consumer of energy in relation to loading. Hence from a passenger-mile standpoint, a fully loaded bus or commuter train consumes the same amount of energy as an empty one. (For buses, this is not strictly true, but for our purposes it is a reasonable assumption.)

The question is how, in one metropolitan region or another, can work-trip travelers be made to shift to these modes? From recent work done in the Chicago region (Transportation Center, 1973) the issue is one of utilization of the transit capacity in those ways that most efficiently link residences in city and suburb to work places in city and suburb. At present, transit provides this accessibility only for work places in the city of Chicago.

VEHICLE SIZE LIMITS

One of the major costs in energy is the fact that about two-thirds of the urban travel by automobile is done in full-size vehicles. Since a full-size vehicle consumes energy at a rate three times that of a subcompact, it is reasonable to examine the energy savings from a policy requiring all urban travel by automobile to be done in small vehicles.

PERSONAL RAPID TRANSIT SYSTEM

A new technology for urban transportation which may offer energy savings is personal rapid transit (PRT). For this evaluation, an electric powered, dual-mode system was used. Basically, this assumed a 1500-pound vehicle driven at a speed of 20 miles per hour no more than 1 mile to a guideway and routed through a network at a speed of 20 miles per hour to within 1 mile of a work site. The vehicle would then be returned to manual control for the final leg to the work site. It was assumed that during manual operation standard storage battery, electric motor propulsion would be used, and that during the time on the guideway, power would be supplied directly from the guideway. It was also assumed that the vehicle would be rubber tired. This is not the most efficient PRT, but it would be the most flexible.

CAR POOLING

Normal automobile occupancy rates for work trips is about 1.3 persons per vehicle. This means that 3/4 of all automobiles during the peak hours contain only the driver. If 1/2 of these vehicles carried 2 persons rather than 1, then the total energy expenditure per passenger-mile of travel would be reduced by 25 per cent, assuming no significant increase in trip length. The basic question is how many urban work trips can use car pooling, and how willing will travelers be to share rides.

SUBSTITUTION OF COMMUNICATION FOR TRANSPORTATION

Another alternative for reducing the petroleum consumption from urban transportation is to make more efficient use of modern communications technology. In theory, much of the nonmanufacturing employment involves information processing and transfer as well as decision-making. Such activities require access to data and analysis capability, and these activities cannot be carried out without bringing large numbers of people together in one place for long periods of time. If the data can be transferred from storage to the individual worker rather than the other way around, much travel would be obviated. Certainly the technology is within the state of the art to permit such a substitution.

Cost Estimates

The cost of each is estimated using the units developed in Chapter 3. Because energy costs are small relative to total internal and external costs of urban transportation, we do this analysis on the basis of net changes from present operation of urban transport. Basically, the total costs of urban transportation, including all the externalities are in the range of $100 to 1000 billion while the oil savings from any of the selected alternatives are in the range of $0.1 to 1.0 billion. Hence the best way to evaluate the effectiveness of policy alternatives for achieving the national policy goal of energy conservation is to look at the overall *changes* that each alternative would bring about.

For the purposes of this example, estimates are presented without proof. They have been derived from available data and in some cases are judgements based on aggregate data when available. In computing the direct costs, annual costs were estimated. For research and development and engineering, it simply was assumed that an average annual cost was required to develop an implementable system. This was to be accomplished in no more than 5 years. For capital investment, it was assumed that all costs would be amortized over 5 years. This is a very high amortization rate, but given rapid obsolescence it is not unreasonable. In sum, the direct costs are annualized based upon a 5-year program period. It is simply assumed that within each alternative innovation would continue and the average cost would be the same for any succeeding 5-year interval. This is important in the feasibility analysis to be discussed later. It should be noted that these assumptions are oversimplifications but are convenient for policy development purposes. It is anticipated that for some policy analyses more detailed data are generated so that detailed costs can be developed.

The results of the policy analyses of the seven alternatives are shown in Table 6.6. The totals represent the net increase or decrease over the existing urban transport process. A negative total reflects a net savings to the society while a positive total represents a net cost to the society. As may be seen, personal rapid transit and land use rearrangements offer the greatest total savings. These are followed by improved traffic flow control and sub-

Table 6.6 Net Annual Costs for Seven Oil Policy Alternatives (in $ Billions)

Cost Category	Land Use Change	Regional Mass Transit	Car Pooling	PRT	Traffic Control	Substitution of Communication for Travel	Smaller Vehicles
Direct[b]							
R&D	0.1	0.3	0.2	0.2	0.3	0.5	0.0
Energy	0.2	0.5	0.2	1.0	1.0	0.5	0.0
Capital investment	0.0	1.5	0.0	1.0	0.8	1.0	0.6
Materials	0.0	0.5	0.0	0.6	0.2	0.2	1.0
Maintenance	0.0	0.2	0.0	0.5	0.1	0.2	0.1
Management	0.2	0.1	0.1	0.1	0.0	0.1	0.1
Labor/enforcement	1.0	0.1	0.1	1.0	0.6	0.5	1.0
Subtotal	+1.5	+3.2	+0.6	+4.4	+3.0	+3.0	+2.8

Indirect							
Energy	−0.8	−0.3	−0.7	−0.8	−0.6	−0.4	−1.3
(barrels per day)	(−210,000)	(−94,300)	(−185,000)	(−218,000)	(−172,000)	(−102,000)	(−374,000)
Time	−10.0	+15.0	+10.0	−15.0	−10.0	−10.0	−3.0
Pollution	−0.2	−0.2	−0.3	−0.6	−0.2	−0.4	−0.1
Safety	−2.0	−1.5	−1.5	−6.0	−4.0	−4.0	+1.0
Natural resources	−2.0	−1.0	−0.5	+1.0	0.0	−1.0	−1.0
Public health	0.0	0.0	0.0	0.0	0.0	0.0	0.0
Land	0.0	0.0	0.0	0.0	0.0	0.0	0.0
Reliability	0.0	+1.0	+1.5	−1.5	−1.0	0.0	0.0
Job opportunity	−1.0	−1.0	+0.5	−1.5	0.0	0.0	+0.5
Shelter quality	0.0	0.0	0.0	0.0	0.0	0.0	0.0
Esthetic quality	0.0	0.0	0.0	0.0	0.0	0.0	0.0
Crime	0.0	0.0	0.0	0.0	0.0	0.0	0.0
Total	−14.5	+15.2	+9.6	−20.0	−12.8	−12.8	−1.1

[a] − = savings to society
+ = cost to society.

[b] Annualized costs assuming 5-year amortizations where appropriate.

stitution of communication for transportation. A shift to smaller vehicles from one-third to two-thirds of the total urban automobile fleet provides a net benefit of approximately $1 billion. Finally, improved conventional transit and car pooling produce a net increase in costs.

Looking at the elements of indirect cost, there are three costs that make up the bulk of the indirect effects. These are time, safety, and natural resource utilization. Time is clearly the dominant one. What this analysis says, in effect, is that methods for reducing oil consumption have their greatest individual social effect on productive time. For five of the alternatives, it is a net gain in productive time that is produced. Secondarily, it is safety that is affected most, and all the alternatives except that of smaller vehicles produce a net reduction in accident costs.

It is interesting to note that as a proportion of the total costs, oil savings (calculated, by the way, on the basis of $10 a barrel by 1973) are a relatively minor contribution. We are talking about a net saving, maximum, of 374,000 barrels of oil a day or $1.37 billion a year. In two aspects this is striking. First, the total impact of oil conservation measures in urban transportation is very great indeed, although the oil savings, in dollar terms, is quite small. Second, the maximum amount of possible oil savings by any of the alternatives is quantitatively very small relative to projected demand. The best conservation alternative provides only a reduction of about 2 to 3 per cent in total demand.

Again, the results of this analysis indicate quite clearly the direct and indirect impact of a policy alternative. The indirect effects are the most significant and are only remotely related to the manifest objective of this particular policy development. By going through the analysis, we are able to identify what these externalities are and hence have an overall insight into the impact of any policy alternative. This is a major objective in policy analysis.

TEMPORAL FEASIBILITY ANALYSIS

From the previous analysis, it is learned that all seven alternatives produce gasoline savings, albeit modest ones. Regional mass transit programs and car pooling generate an overall increased annual cost to the society while the remaining five provide an overall benefit. Since the purpose of policy analysis is to reject no alternatives nor to weight any of them, the fact of the estimated variation is simply noted. The policy-maker has the responsibility for the interpretation of the analysis. However, the policy analysis is not complete. It must also answer the question of how feasible it may be to implement any of the alternatives. To complete the policy analysis, the policy-maker must be provided estimates of the feasibility of implementing each of the policy alternatives.

Time Requirements

Temporal feasibility is the time rate at which any policy alternative may be implemented and hence the rate at which the benefits of any policy decision may accrue to the society. We are concerned in this analysis with the estimation of the delays that occur from a point of decision to full implementation. In the first instance our concern is simply with the time required to obtain the estimated oil savings. In the second we are concerned with the cumulative benefits and costs over time of the implementation of the seven policy alternatives.

To implement any of the seven urban oil conservation measures, each of five time-consuming stages must be completed. These are as follows:

1. Research and development
2. Engineering and production
3. Institutional arrangements
4. Organization and manpower
5. Installation

The first two have some flexibility in the sense that money and manpower may be substituted for time. Although there is a minimum time required to develop a new system, the maximum is simply determined by resource allocation. Similarly, how the public sector chooses to stimulate the production of any new systems is determined by public policy decisions. For example, during the second World War, the United States government subsidized the development and construction of synthetic rubber production facilities which were operated by the major rubber manufacturers.

The third and fourth stages are far less easy to determine than the first two. In the American system, state and local government have the ultimate control over implementation. The political and administrative structure and its adaptability determine whether or how long it will take to implement any system, social or technological. The more extensive the social or institutional change a new system requires, the longer it will take to implement. Similarly, the existence and number of private entities available for managing the implementation is also a crucial determinant of the speed with which a policy alternative may be implemented. Unless there is sufficient construction or administrative capability, then the rate of implementation is bounded by this limitation.

Finally, the rate of installation is always limited. The characteristic of the system and the administrative procedure for control of the implementation process determine ultimately the rate of installation. The construction of a highway, a waste water collection system, or a building is always a time-consuming operation which can go on no faster than design and structural criteria permit. For purposes of the analysis of oil conservation policy alternatives, we estimate the time for each of the elements. They are the minimums that may be expected, assuming the direct cost investments defined in the previous analysis. We evaluate each alternative in turn and plot them on the same time frame.

LAND USE REARRANGEMENT

We suppose that both social rewards and sanctions are used to stimulate relocation so that people move closer to their places of employment. The positive incentives may be tax benefits, government loans, renewal programs, and such. The negative incentives might be travel taxes, fuel pricing policies, tax penalities, and such. Such a program would be expected to stimulate people slowly to choose residences nearer to their places of employment. Given this kind of relocation policy and program, we can identify the time phasing for this alternative. There appear to be three stages:

1. Promulgation of incentive policies
2. Identification of housing stock and administrative mechanisms to facilitate relocation
3. Turnover of housing and shift to shorter work trips

We assume that the first stage requires 2 years to fully complete. The second stage is assumed to require an additional 2 years to make functional on a nationwide basis. Finally, with this forced relocation program, we would expect a rate of change of the order of 10 per cent a year maximum and hence a 10 per cent reduction in oil expenditure per year up to the maximum of 210,000 barrels of oil per day. This is plotted in Figure 6.1.

REGIONAL MASS TRANSIT

For most major metropolitan areas an effective public transportation system requires linkage of suburban residences to jobs in the suburbs as well as to central city jobs. Similarly, it must link city residences to suburban as well as city employment sites. Basically, such a system will be fixed schedule and fixed routing systems. Regionalization of bus and rail would be accomplished by organization of interchange locations that effectively interconnect radial, circumferential, and collector distributor routes. Such a system efficiently operated would provide maximum connectivity for linking residences and employment sites within any metropolitan region.

Three elements determine the rate at which oil savings may be obtained. One is the institutional arrangement necessary to create a regional authority operating across jurisdictional boundaries and having the financial capability to obtain essential capital equipment. Second is the organizational capacity to effectively operate a complex regional network. Third is the productive capacity to supply the capital equipment needed to operate a regional system. We estimate the time for institutional arrangements at 2 years. The time for research and development needed to provide the operating capability and the administrative organization is approximately 2 years. Implementation of the complete operational system takes 5 years. This too is plotted in Figure 6.1.

CAR POOLING

To increase car occupancy, two major steps are taken. One is to provide positive and negative incentives to make car pooling more attractive. The second requirement provides a communication capability sufficient to link people having common areas of residence and common employment locations. Also, of course, there is to be an administrative mechanism for operating the system. Basically, we would estimate the minimum time necessary to place a car pooling system in operation to be approximately 3 years. It would then take approximately 3 years to obtain maximum participation. The time function is also shown in Figure 6.1.

PERSONAL RAPID TRANSIT (PRT)

This is a wholly automated system of small-tracked vehicles. The single mode system used for this evaluation is conceived as a dense grid network of guideways with vehicles automatically routed and operating in a demand actuated mode. This is, of course, a very high technology transport system requiring significant amounts of research and development and engineering. Major new institutional arrangements have to be developed to implement such a system. Finally, wholly new production systems have to be created to produce sufficient quantities of the system components. We estimate that a minimum of 5 years is required to develop an operational system, and an additional 5 years is needed to create the production and implementation capability. Since we are talking about 40,000 miles of guideway

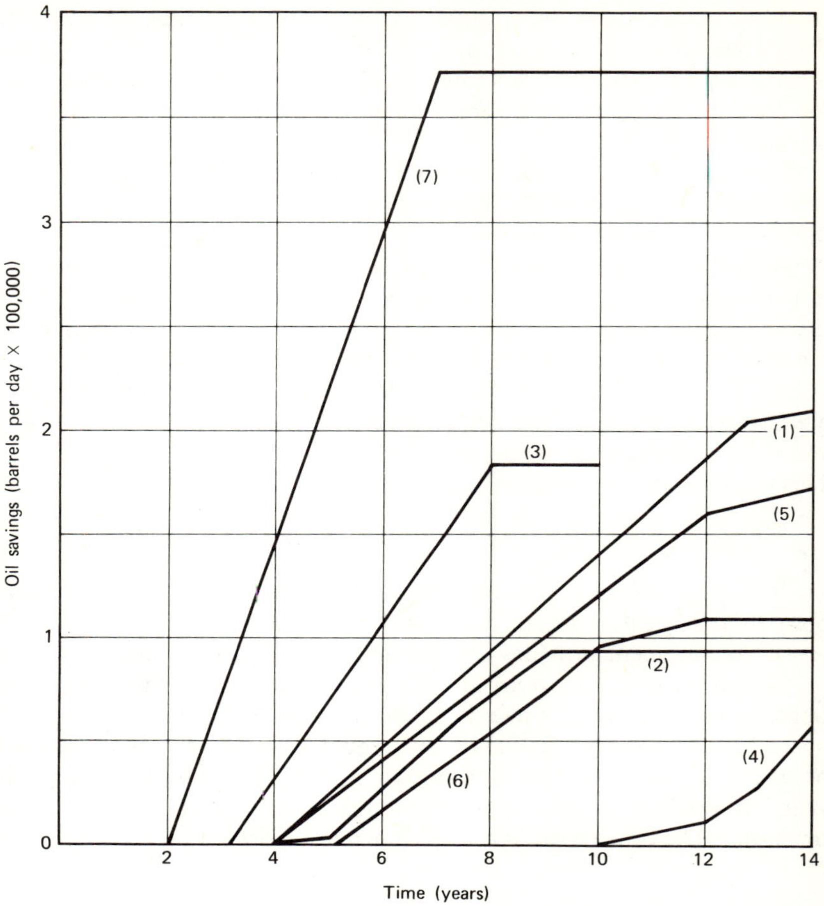

Figure 6.1. Time required to obtain oil savings for seven alternative conservation measures. (1) Land use change; (2) regional mass transit; (3) car pooling; (4) PRT; (5) traffic control; (6) substitution of communication; (7) smaller vehicles.

and around four million vehicles, the total installation time is approximately 10 years. With these estimates, we can determine the time rate of savings. This also is shown in Figure 6.1.

TRAFFIC CONTROL TECHNOLOGY

Electronic traffic control systems of various types have been developed that can significantly increase the carrying capacities of individual streets as well as highway networks. Such systems require devices located both in roadways and in vehicles. All are capable of being added to existing highways as well as to vehicles presently on the road. Considerable research development and engineering would be required, as would production and installation capability. Expansion of electronic maintenance technical skills would also have to be developed for operating these systems. We estimate that 3 years would be required to develop and begin producing the appropriate technology. It is then estimated that it would take a total of 10 years to install the components on the urban arterial and freeway networks and all vehicles using these urban highways. Since the gains in oil savings would not begin to accrue until 20 per cent of the vehicles and highways were instrumented, the rate of growth of savings would be nonlinear. The basic function is again shown in Figure 6.1.

SUBSTITUTION OF COMMUNICATION FOR TRANSPORTATION

To reduce travel for employment devoted to administrative, information handling, and executive functions, a distributed network of communications centers has to be provided. Such systems require significantly more complex and sophisticated technology than currently available. This includes transmission, display, storage, and retrieval capability that exists primarily in the military. Obviously, such a program requires major changes in institutional organization now predicated on face-to-face contact.

Finally, the capitalization and construction of the necessary facilities must be provided. Of all of these steps the second is the one that requires the largest increment in time. The technology and production capability can be obtained within 3 years. The time required for organizational changes probably cannot be less than 5 years. Thus first implementation cannot occur in less than 10 years. The time of obtaining the maximum oil savings is shown in Figure 6.1.

MORE UTILIZATION OF SMALLER AUTOMOBILES

Of all the policy alternatives examined, the easiest to implement technically is increased use of smaller vehicles. Essentially, what is required is increased production facilities and an incentive program to cause automobile buyers to select more energy-efficient vehicles. A variety of positive and negative incentives can be used, from outright subsidies to tax or pricing policies. The latter have been used successfully in Europe to limit sales of large vehicles. Thus the key issue is not technological but social, in the sense of shifting preferences from larger to smaller vehicles. We may estimate the production lead time at 2 years assuming no attempt is made to design an original urban vehicle. The second stage, inducing the shift, takes longer but should be accomplished within 5 years, assuming a conventional half-life of a vehicle. Thus for two-thirds of the vehicles in use, a shift from two-thirds full size vehicles down to one-third could be accomplished in a total of 7 years. This is plotted in Figure 6.1.

Other Benefits and Costs

In these analyses of feasibility we have looked only at the specific benefit of oil savings. It is important in the context of a complete policy analysis to provide a temporal estimation of savings or costs to the society as a whole. As the previous analysis has shown (Table 6.6), the seven alternatives generate significant external benefits and costs. From an overall societal standpoint, this estimate is determined in terms of feasibility. Given the time estimates developed above, the the estimates of the direct costs necessary to produce and install this system, it is possible to determine the cumulative costs to the society of implementing each of the alternatives. These are plotted in Figure 6.2 for each of the seven alternatives.

There are clear differences in the results of the two analyses. Because of delays to implementation, there are net costs to the society as a whole as resources are expended to create, produce, and organize the implementation of any of the alternatives. For mass transit alternative and the car pooling alternative, these costs increase as they are implemented. This is true because they reduce mobility by mainly increasing travel time.

Five alternatives produce net benefits to the society over the long run. However, they do so at differential rates. These rates of return are determined by three factors; the costs of implementation, the total benefit of the alternative, and the delay to implementation. The last is seen in the personal rapid transit alternative, curve 4 in Figure 6.2. Because it is estimated to take 8 to 10 years to develop and begin implementing, it is not until 17.5 years after a decision to begin the program that a net return to the society occurs. It is well to point out that the slope of this benefit function is greater than any of the others and eventually the net return to the society would exceed land use rearrangements (curve 1), substitution of communication (curve 6), or improved traffic control (curve 5). However, this crossover point would not occur until 40 to 50 years had passed.

Conversely, relocation of housing near to employment has a low development cost and a fairly high rate of return. Thus it produces a net benefit to the society after only 7 years from its initiation.

It is also interesting to note that alternative 7, switching to smaller automobiles, generates the greatest oil savings in the shortest period of time.

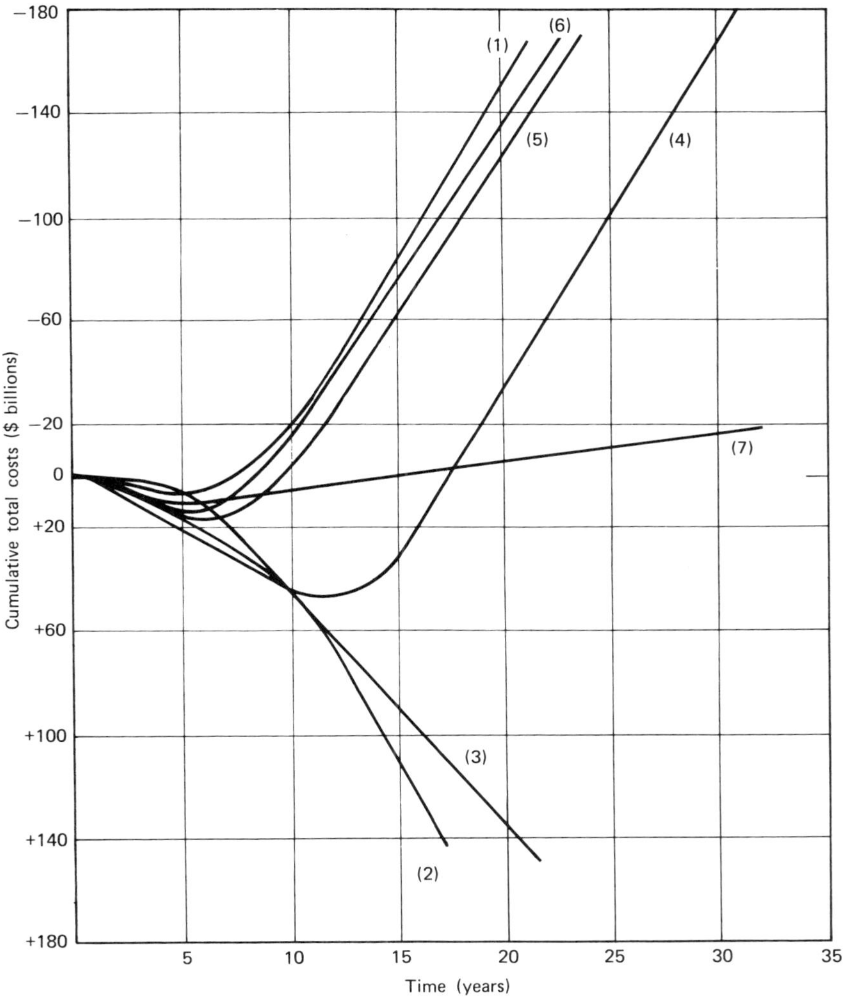

Figure 6.2. Cumulative societal costs of seven oil conservation alternatives. (1) Land use change; (2) regional mass transit; (3) car pooling; (4) PRT; (5) traffic control; (6) substitution of communication; (7) smaller vehicles.

However, it does not produce a net benefit to society until 15 years after its initiation. Compared to the other benefit producing alternatives, it never generates a large net return to the society.

Another point of importance that emerges from these analyses is the long lead times that are necessary before the benefits from any of the alternatives begin. In Figure 6.2, the point of inflection of each cost curve defines when implementation is complete. This is also very clear from the curves shown in Figure 6.1. Thus for the seven alternatives the minimum full implementation time is around seven years (transition to smaller vehicles). The maximum is around 20 years for PRT.

These analyses also demonstrate the importance of advance planning. In the context of this problem, a 10- to 15-year lead time was provided. This was the time interval before an energy crisis would emerge if nothing were done. All the alternatives except PRT are able to meet that time frame. If, however, the society were to defer decisions for 10 or even 5 years, the country would be faced with an energy crisis about which little could be done in the short run. Consequently, the society would have to suffer several years of social and economic dislocation. We can look at our societal performance criteria and estimate what the total annual societal costs would be for an energy shortage of any magnitude. A crude estimate of costs suggests they would be of the order of $15 to 20 billion annually for a 10 to 20 per cent shortfall.

To summarize the results of the feasibility analysis, we have been able to estimate the time and total return on any investment in urban transportation energy conservation alternatives. We find that five of seven alternatives provide a net benefit to the society while two produce a net cost. This is true even though all seven do produce energy savings. We also find that the most cost effective alternative, PRT, appears to take the longest time to implement. The alternate which produces the greatest oil savings in the shortest time, increased use of small cars, is not the most cost effective for the long run, from an overall societal standpoint. In essence, our analysis indicates that the time frame within which action is desired significantly influences cost effectiveness.

QUALITATIVE FEASIBILITY ANALYSIS

The previous analysis provides time cost estimates for the seven alternatives. Such analysis is not sufficient for policy decision-making. We need also to estimate five other dimensions to provide a complete evaluation of the feasibility of our seven alternatives. These feasibility dimensions were defined in Chapter 3 and are listed below:

1. Technological
2. Social
3. Economic
4. Political
5. Temporal

As has been discussed, there is no way to define quantitatively these dimensions because they are largely judgemental in nature. Consequently, we must use a scaling method that provides a reasonable measure of subjective judgement.

We apply the technique developed in Chapter 3 to each of our seven alternatives. To do this, we provide the description of each alternative and its desired performance to an appropriate sample of people. (By appropriate we mean those groups within the society knowledgeable of the dimension we wish to scale. For technological feasibility this obviously refers to scientists and engineers. For political feasibility this refers to local and state as well as national politicians. Social feasibility may be estimated from a sample of citizens representing a cross section of the society impacted by the policy alternative.)

Let us suppose we ask the appropriate groups to evaluate the feasibility in each dimension on the nine-point scale and further assume a random representative sample of between 100 and 1000 individuals. For each group and for each dimension we obtain a scale score which may be converted to an interval scale using any of several of the Thurstonian models. For our purposes, we assume that the nine-point scale is a true interval scale with hence a unit variance. Within this set of assumptions we may derive a mean value for each feasibility element for each policy alternative. This is shown in Table 6.7. On the assumption of dimensional independence, homogeniety

Table 6.7 Feasibility Estimates of Policy Alternatives

	Policy Alternative						
Feasibility Dimension	Land Use Change	Regional Mass Transit	Car Pooling	PRT	Traffic Control	Substitution of Communication	Smaller Vehicles
Technological	6	5	6	3	4	4	8
Social	2	4	3	7	7	4	3
Economic	3	5	7	4	6	5	7
Political	2	5	3	5	7	4	3
Temporal	5	5	6	2	5	4	7
TOTAL	18	24	25	21	29	21	28

of variance, and true interval scale, the columns may be summed and the totals represent the overall evaluation of the feasibility of each alternative. We can see that the traffic control alternative, the most feasible of the alternatives, is ten units more preferred than land use change. In general, the traffic control and switch to small cars appear to be the two most preferred of the seven. Significantly lower are car pooling and regional mass transit.

It is now instructive to look at all seven alternatives in the four dimensions of importance to a policy decision. These are: (*a*) oil savings; (*b*) societal costs; (*c*) time to full implementation; and (*d*) judgmental feasibility. The numerical values for each are shown in Table 6.8, all in their own units. This summary table is very illuminating. All of the evaluation dimensions are virtually independent. Those alternatives that provide the largest societal benefit in dollar terms are not related to their subjective feasibility. As a matter of fact, a simple rank correlation of the two dimensions yield a rho of −0.4 which, although it is not statistically significant, is indicative of how differently the dimensions are rated. It is important to note that this independence is a positive finding. Effectively, it indicates

Table 6.8 Summary of Evaluation of Seven Oil Conservation Policy Alternatives

	Policy Alternatives						
Evaluation Dimension	Land Use Change	Regional Mass Transit	Car Pooling	PRT	Traffic Control	Substitution of Communication	Smaller Vehicles
Oil savings (b/d × 100,000)	2.1	0.9	1.9	2.2	1.7	1.0	3.7
Societal costs ($ billions)	−14.5	+15.2	+9.6	−20.0	−12.8	−12.7	−1.1
Time to implement (years)	13	9	8	18	14	10	7
Feasibility rating	18	24	25	21	29	21	28

that our evaluation dimensions are unique factors, each measuring a different element of social concern. In complex social issues, this is what we should expect. Only rarely and only in simple issues would we expect to find a convergence of the evaluation on a single policy alternative. This result also shows that there are unique domains, all of which must be considered in policy-making. A technical or economic evaluation or decision model is simply not sufficient if policy-making is to be responsive socially.

Another important result of our analysis is that what may be technically and economically worthwhile is not necessarily socially acceptable. It is well to note that the selection of a policy alternative can and has been made on the basis of any one of the four evaluation dimensions. Obviously, if the largest oil savings in the shortest possible time is the sole concern of an advocate, he justifies forcing a switch to smaller vehicles. However, he overlooks the social and political disruptions such a policy might incur. Similarly, he overlooks the larger and longer term benefits of other alternatives.

Finally, Table 6.8 shows very clearly the policy-maker's dilemma. Conscientious and socially responsive policy-making must consider and rationalize the difference among the four evaluative dimensions. The process must weight them, especially when there are significant conflicts, and must select a policy alternative that best satisfies social goals at least cost, political and social as well as economic. In the end, the policy-maker is the one responsible to the society for making the trade-offs necessary to a policy decision.

The policy decision problem should not be underestimated, as it often is by technicians in particular. The trade-offs that have to be made are difficult ones and are very clear in the present case. The short-term versus long-term benefits and costs are obvious. To choose an alternative that satisfies a social goal only in the short run may be (and is in this case) a simple problem. However, to consider or select an alternative with a long-term benefit but no short-run payoff has social risks. If such is selected and it takes longer to develop and implement than estimated, a crisis will occur, one that could have been avoided. Selecting an alternative that is cost effective but socially unacceptable may preclude its ever being implemented without a long period of social conflict. This appears exactly to be the fate of urban expressways.

In essence, the policy analysis process is complete with the feasibility analysis. We have gone through the process and generated four classes of information essential to policy decision-making. Without any biases imposed by the analyst, the results contained in Tables 6.6 through 6.8 and Figures 6.1 and 6.2 can be provided for the decision-maker. What he does and how he uses the analysis must be independent of the analyst. Except insofar as the decision-maker may question and/or wish to change one or more of the estimated values, the policy has reached a decision level.

The policy-maker may wish to weight any of the direct or indirect costs. For example, time cost may not be considered of equal dollar importance for this issue as the other elements. Thus he would apply a weight having a value of less than one to this row in the cost savings matrix. The column totals could then be recomputed and a different result obtained. Or the policy-maker may consider certain of the evaluation dimensions of greater or less importance and hence would weight them accordingly. For example, subjective feasibility might be considered more important than the other three and shift the decision toward one alternative. In the end, the policy analysis procedure simply provides a method for combining a variety of basic elements of social policy-making in a common framework. Its value is that it provides comprehensive information and in a way that allows the decision-maker to understand and communicate the implications of his choice process. In this sense a good policy analysis is neutral and unbiased; it is a tool and a resource for the policy-maker.

CONCLUDING OBSERVATIONS

This analysis was carried through to a level of completion for only urban transportation. The task of analyzing rural and air transport can be likewise developed as well as facing the issue of whether more than one of the alternatives evaluated can be developed and implemented simultaneously. Obviously, if this can be done, the potential benefits in all dimensions may be markedly increased or projected costs minimized.

An element of analysis omitted here is the concept of hierarchical implementation. Clearly, PRT might well provide a long-term solution to the urban transport energy problem, but the uncertainty inherent in its development might well preclude its selection as the single policy alternative. However, it is reasonable to consider a policy for urban transport, for example, that pursued both the switch to smaller automobiles and a vigorous research and development program in PRT. If as both programs progressed it was found that PRT could be implemented sooner than expected and at a faster rate than the initial analysis indicated, then the program of inducing a switch to smaller vehicles could be reduced or abandoned in favor of the PRT. It is reasonable within the policy analysis process to evaluate this class of possibility and with adequate information flow to update those analyses on a regular basis. Thus the process permits multiple programming rather than single alternative decision making thereby increasing the likelihood of meeting desired societal goals. The short-run price of such programming is higher investment costs and lower short-run returns on that investment.

Finally, the time frame issue for decision-making is an essential one. Beyond the "results now" requirement, which is an occupational disease of public policy-makers, the issue of timing is a critical one. On the one hand, the time frame for decision-making must be long enough to permit the implementation of rational and cost effective programs. Progress is slowed when the overriding concern is for immediate crisis response. Policy development is the political analysis by which a society can achieve its goals in an orderly and timely fashion. Time is the essence of that process. If a minimum analysis time does not exist, the society is essentially out of control; that is, progress is random, restricted, or non-existent.

Conversely, there is an upper limit to the policy analysis time horizon. Social needs and goals grow and change. What is needed in the near future may not be needed in the far future. Adopting goals accepted by a present generation—to be met a generation later—is risky at best. Furthermore, the means to attain societal goals are evolving very rapidly. What is technologically feasible at present will, in all likelihood, be obsolete in a generation. Thus the time frame for socially useful policy should never extend beyond a human generation, and probably half a generation would be better. This suggests that the policy development process should operate on a forward analysis time of 5 to 10 years and policy program life of 15 to 20 years. It is remarkable how few institutions, public ones in particular, can conceive of their life span as so short. Finally, if policy development is to operate in this size time window, then it is clear that the level of information and the sophistication of the policy analysis process have to be far greater than anything now available to any public agency outside of the military.

PART 3

AN EVOLVING PROCESS

Public policy development in the early 1970s was not established as an organized process as has been proposed in Parts 1 and 2. Far from being a systematic process such as administration, planning, or design, the analyses that were used as a basis for policy decisions were either very simple and intuitive or advocacy studies for or against a proposed course of action.

The authors of this book are convinced that the easy route of intuitive and self-serving policy decisions is slowly but surely being closed. Political decisions are receiving much more careful scrutiny, and both sides in a controversy are more likely to be heard. Since each and every course of action will find both support and resistance, and since a generally accepted approach to policy development is lacking, the decisions will be unnecessarily erratic and contradictory. Progress of any type toward any goal will be delayed.

Such conditions were evident in the federal government in the 1960s. First, strong support for the development of nuclear energy was removed, then laws to remove environmental pollution were enacted, and finally a major program in support of energy was initiated in the 1970s. These actions plus others in support of safety and health resulted in a cost of living increase—at least partly due to the change in quality of the products, energy costs, and environmental protection.

Politicians were being blamed for policy decisions that appeared overly sensitive to the attitude of the voters or of a powerful lobby. But erratic and short-sighted decisions on resource allocation cannot be blamed entirely on the political process. The technical community was not as yet addressing its efforts to the policy analyses that were needed. Rather, they were continuing to argue for their own interests (on technical matters) and were joining citizen groups to fulfill their political interests. And we believe that this condition can and will be changed.

During the closing years of the twentieth century, the creation of public policy that is responsive to the need for cultural progress is very likely to become more sophisticated—more capable of utilizing science and technology from all areas of formalized knowledge. Pressures for progress, without regression, will require innovative solutions to many types of problems. The demands for energy, communication, and transportation opposed to the demands for preservation of resources, reduction of pollution, and greater personal safety will certainly not be resolved without innovative solutions and creative policy development. A second major thesis of this book is that technology can only be controlled by establishing explicit goals and equally explicit constraints.

The chapters of Part 3 are directed to the problem of how to evolve a more effective policy development process. In Chapter 7, a history of man is presented, leading to a suggested series of cultural changes that are likely to occur during the last quarter of the twentieth century. It is hoped that this chapter will assist in developing an understanding of the issues that will need to be resolved.

Chapter 8 presents both an academic and a practical view of the political process. Certainly a positive link between the technical and political processes requires that the real—not an ideal or imaginary—world of politics be understood by technologists and engineers. Chapter 9 provides an overview of policy development and adds some additional principles for applying resource allocation at various levels.

CHAPTER 7

THE SOCIAL MOTIVATION OF POLICY DEVELOPMENT

It is the underlying assumption of this book that the nature of man and of his social institutions has been and is in the process of transition from one world view to another. It thus becomes clear that the means by which a society, now largely through government, makes use of technology, and the criteria by which it applies to engineering works are inadequate for the changing goals of man. To a large extent these inadequacies derive from a basic failure of technologists on the one hand and public policy-makers on the other to grasp clearly the changing social priorities and to adapt science and technology to the new goals of man.

If wise and adaptive application of science and technology is to be used as a tool of social development in a new era, engineers and technologists must become more fully aware of the changing society which they serve and the goals for which their skills are prerequisite if those social goals are to be met. Government, at all levels, which makes the resource allocation decisions for engineering works in light of social objectives, has the same task. In addition, however, the government must make far more sophisticated and rational applications of science and technology than it had up to the 1970s. The purpose of this chapter, therefore, is to do three things: the first, is to explore the new world view of man. The second, is to develop a rationale for the relation of public policy to science and technology. The third, is to evaluate certain current public works policy decisions in terms applicable to emerging issues requiring technological innovation.

PROCESS OF VALUE CHANGE

To understand the necessity for an increased sophistication in utilizing scientific capability, it is first necessary to examine the changes in society. The forces at work in developed societies markedly shifted the priorities and, more importantly, the values that determine decision-making at the private, as well as public, level. The issue is farily simple: The last half of the twentieth century has seen a movement away from primary consideration of physical welfare to social welfare.

To the engineer, this shift has led to the basic obsolescence of traditional engineering decision-making which was predicated on economic evaluation or, more exactly, economic justification. Similarly, public policy decisions relative to public works have also been rationalized on the basis of economic resource allocation to obtain physical security. One need look only at major programs of the federal government to see this emphasis. The flood control programs of the Corps of Engineers, the federal-aid highway program, and national defense are massive investment programs whose justification is either security against the unpredictable or the economic development of a region. Poverty, protection, and economic exploitation have been the dominant motives underlying the allocation of resources and engineering decisions.

It is essential to understand the roots of change of the value structure in Western society, for new criteria are emerging for setting priorities and for allocating resources. The basis for exploring these new criteria has, to a significant degree, developed out of advances in the behavioral sciences in the last 25 years. These areas—limited though they may be—provide a means of understanding the process of value change and an insight into the criteria for setting and evaluating public policy decisions.

The view of man that has emerged is complex in detail, but quite simple in its major form. The essential thrust has been a shift away from concentration on the *explicit behavior* of man to what *motivates that behavior*. This is a deceptively simple statement, but its implications create a direct confrontation with traditional social values. Before exploring these conflicts, it is useful to explain this view of man more fully.

Man may be described as a needfulfilling organism. A need, in its most basic definition, is a physiological or psychological deficiency which the individual must eliminate. In the lowest order of such needs are bodily ones such as hunger, thirst, and metabolic requirements. As these deficiencies arise, they will generate tension in the individual. These need states may be consciously recognized, for example, hunger pangs, or may act unconsciously, for example, generalized tension. In either case, they increase the activity level of the individual and initiate a process of goal-seeking behavior. The individual so motivated begins to act on his environment to obtain those objects that will satisfy the deficiency and the tension it creates. At the simplest, man—like all organisms—may be conceived of as an equilibrium system which, when driven out of equilibrium, acts on the environment in such a way as to return to equilibrium with a minimum expenditure of effort.

The important aspect of this process is that the observable overt behavior is itself significant only because it implies a motivation or a need. Essentially, the behavior is a symptom of the disequilibrium state of the individual. What is significant is that it is the underlying need structure that explains or determines behavior. It is only on this basis that the rationality of behavior becomes both understandable and ultimately determinate.

What has been said up to this point is readily applicable and valid for all animals, including man. In this sense, man's evolutionary connection to all living organisms is clear and consistent. However, man has three properties that distinguish him clearly from the rest of the animal kingdom that resulted in his emergence as the planet's dominant life form.

The first is the development in man of long-term and plastic memory. There is abundant evidence that man can store and recall larger amounts of information than any other species. Not only can he respond to experience, to stimuli, he can remember them and use them repeatedly. More importantly, he can manipulate these experiences and stimuli abstractly. With this capacity, he can evaluate the experience, compare it with others, and alter it (ideation) to create new relations. With the development of symbolic notations, for example, language and number systems, the range of manipulation of experience becomes almost indefinitely large. It is this ideational process that is the unique aspect of all human learning. Although all animals learn by simple conditioning, as does man, none have the magnitude and diversity of association of stimuli that man has nor the capacity to learn through abstracts. Finally, none has the long-term storage and recall. It is through these processes that man has been able to interpret and hence control his environment and, incidentally, control or be controlled by other men.

A second property of man is his highly developed capacity to operate on his environment. Again, most animals have some capacity to act on their surroundings, but in none has the capability evolved as in man. In part, this is because of the memory capacity of man and in part it is because man has evolved with a highly developed manual dexterity. The simple articulation of fingers, hands, and arms is more highly developed in man than any other animal. More importantly, the fine control of these elements is so great that man can direct their action with remarkable precision, and they are coupled with a sensory system of remarkable sensitivity. For example, the human eye is capable of detecting light energy of the order of one photon and can detect movement as slow as 10^{-4} radians per second. With these capacities and a highly developed representational process, it is little wonder that man is called the tool maker.

This manipulative capacity of man has been his primary means of dealing with his environment. Unlike most other animals, he can consciously and in a directed fashion modify natural materials to construct a man-centered environment. He can thereby on the one hand moderate, if not overcome, physical variations in the environment and, on the other, harness natural forces. It is the development of these capabilities that permitted man to create agriculture and hence amplify his energy by an order of magnitude. If it is recognized that man operates to satisfy his needs with a minimum expenditure of energy, it can be seen that a shift from a hunting-gathering activity to a controlled agriculture was a quantum improvement in energy utilization. In fact, until well into the twentieth century, every discovery that offered such energy reduction was quickly put to use and universally rewarded.

A third property of man, which is of fundamental importance, is his complexity. The degrees of freedom in human behavior boggles the mind. Not only is this true in terms of overt action, but also it is true in terms of variation and change within any one individual. Within the perfectly general and relatively simple process of need fulfillment, the alternative means and ends for attaining goals are almost infinite. It is for this reason that traditional taxonomies of human behavior and characteristics are futile. These are outputs of a process and are, therefore, idiosyncratic or situation specific. To try to classify people on this basis leads to gross oversimplications at best and dehumanization at worst. The fact is, the only way to understand human behavior is in terms of motivation. How man actualizes himself in observable action is so variable and hence unpredictable as to be a useless frame of reference. It may be observed that this flies in the face of conventional wisdom and a very long Western tradition. Men have always been viewed as the sum of a set of behavioral traits generally on scales of semantic opposites such as strong-weak, bravecowardly, bright-dull, and cooperative-competitive. In general, the conventional wisdom has treated man mechanistically and created its explanations in terms of man's overt behavior.

Modern conceptions of human behavior deal largely with motivations—what initiates or drives or directs behavior. At most, observed behavior is a starting point from which it is possible to work back to underlying needs. It may be noted that operating on such principles leads to a difficult conventional problem: It makes it apparently impossible to generalize about people. Since the Western world, at least, has built its social and political philosophies on the basis of certain general assumptions about man's behavior, such a change threatens the basic fabric of traditional social order. It may be added that it threatens the basic assumptions upon which engineering practice is predicated. A mechanistic view of man underlies almost all concepts of civil public works and is most manifest in benefit-cost evaluations. This most fundamental value of engineering presumes that man and his society act to derive the maximum economic return for a public works investment. To do anything else is wasteful. It is unfortunate, perhaps, and even painful, for traditional engineers to discover that resistance to these values has grown to the point of preventing construction of public works that have high benefit/cost ratios.

These cases are actually symptoms of a basic change in priorities and in man's view of himself. This shift emerges out of the changing structure of needs—with the American society, at least.

GOALS IN TERMS OF HUMAN NEEDS

The complexities described above require a fuller examination of the concept of needs. Up to this point, only a definition of a need has been given. The structure of needs and their organizing properties have not been described. It is clear that needs themselves are classifiable into a relatively small set. For our purposes, they may be categorized in three groups: (1) survival needs, (2) security needs, and (3) ego, self, or being needs.

Survival Needs

Survival needs are those primary motives generally physiologically based. These include all the bodily deficiencies. They are not restricted to conscious needs such as hunger and thirst, but also the whole class of metabolic deficiencies such as avitaminosis and temperature imbalances, all of which motivate the organism to seek agents within the environment which eliminate these deficiencies.

Needs in this class are always first-order needs. Unless and until these deficiencies are met, the organism is in a state of tension and disequilibrium. All his energies, of necessity, must be devoted to the search for those agents that satisfy the deficiencies. Valuable are those objects, processes, or behaviors that adaptively provide satisfaction of survival needs. It is obvious why man, upon discovery, rapidly shifted from a hunting-gathering culture to an agricultural one. It is equally obvious why man had to develop settlements around a stable water supply, and why persons were rewarded who, by invention, increased the likelihood of meeting survival needs. In sum, man's existence and his social organization were motivated by the need for survival, and they concentrated on satisfying survival needs.

It is difficult in the latter half of the twentieth century, especially for Western society, to grasp the significance of survival needs. Yet prior to this century, man was utterly helpless in the face of most physical deprivation. Diseases had no cures (nor was it generally known that they had causes). Floods were unpredictable and largely uncontrollable until late in the nineteenth century. Crop failures due to insects were uncontrollable, leaving famine in their wake. The marginality of human existence was the central condition and the central force directing individual and group behavior. What men did and how they did it was motivated by the overwhelming concern with survival needs.

It is well to reflect on man's values and the attitude that directed his social organization. Values as used in this context may be defined as the priorities man sets for adaptive behavior. In a survival society, physical strength and bravery were of high value. These statements are also, of course, true for the few remaining survival societies in the late twentieth century. Longevity had to be of high value, also, if for no other reason than that in an empirical world, experience, qua age, provides increased knowledge, lore, and technique. Veneration of the old is a very general social value among survival societies, although it is by no means universal. One would expect to find that the main concerns would be focused on, and the main relationships dependent on, the acquisition and maintenance of survival goods. It is consumable goods or their symbolic surrogates which are prized and protected. It is not surprising, therefore, that dowries in survival societies were cattle and foodstuffs. In sum, what had value were those qualities and those objects that increased or enhanced the satisfaction of survival needs.

Social organization was similarly simple. Generally, survival societies were tribal with a single absolute leader. The behavior of members was dictated by the marginal nature of existence, and the conflicts which a chief had to adjudicate related to issues of material possession. Order was imposed by direct action by or through the tribal leaders and by symbolic devices such as taboos. (Taboos are simply behaviors which cause punishment to be meted out.) Interpersonal relations were generally very direct and involve simple physical acts of competition or cooperation. In general, people were objects to be used as a means to meet or sustain survival needs. People had little value other than this. Chattel is probably the focal concept in the functioning of survival societies. Most, though not all, such societies did not have one-to-one relationships. They were usually polygamous, and children were not seen as uniquely belonging to parents. In essence, survival societies operate on simple utilitarian principles.

Finally, most survival societies were nonrational in their approach to life. Nonrational does not mean irrational or thoughtless. Rather it means acting out or behaving on the basis of direct feeling. This occurs in survival societies for two simple reasons. One is that, operating at marginal subsistence levels, man has very limited capacity for emotional control. In those states, and very probably for physiological reasons, his management of complexity is as limited as is his capacity for extended abstract thinking. Under these conditions, he is capable of responding in only the most direct fashion to emotional stress. It is probably for these reasons that social controls are exerted by blunt and usually brutal means. It is, in fact, the only methods that reach the individual and hence cause either learning or social response. Cruel and inhuman punishment is a more recent and novel social concept which in any of its applications would be wholly meaningless and useless in controlling human behavior in survival cultures.

A second reason for nonrational behavior was that man's knowledge and understanding of the world were so limited. Death, disease, and deprivation were everywhere and occurred inexplicably. Knowledge gained was in someone's head and transmitted largely by word of mouth. Learning was largely by direct experience from which the range of generalization was narrow at best. The result was a society bounded by grave and unpredictable threats. Its central concern was to seek protection from those threats. Rationality is not particularly salient under these circumstances. The last thing these cultures sought or rewarded were philosophers, scientists, or poets. Fortunately, for the species, if not the tribe, enough of these maladaptive types were produced to ultimately inch the race along.

With or without the rationalists, man did learn to manage his environment more effectively. Discovering agriculture, learning farming skills, and developing physical technology led to a quantum improvement in human subsistence capability. In the process a new set of needs emerged, as did a new social order.

Security Needs

The second class of human needs may be called security needs. At one level these involve several aspects of physical security. At another level, they involve ego or self needs. Again, the structure of this set is based on deficiencies within the individual which he is trying to satisfy. The major distinctions between these and survival needs is that security motives are psychologically based more than they are physiologically based. And it is at this point that the evolutionary process takes a divergent path from other animals. At the survival need level, the similarity among members of the animal kingdom is quite striking. What distinguishes man from lower animals is his repertoire of response to those needs and his capacity to utilize, in quite formal ways, other members of the species in cooperative behavior to obtain need satisfaction.

Security needs, however, reflect increasingly and more uniquely human processes. Man, with a full stomach, and the ready availability of more food, is still not in equilibrium, or at ease. Tension and restlessness remain. Anxiety rather than physical threat emerges. Moreover, he becomes aware of a different class of feelings: a sense of achievement, differentiation between winning and losing, and growth in his sense of himself as distinguished from others. To a significant degree, security needs emerge from within the individual and are projected onto the physical and social environment where they become objects needed or threats to be avoided or disposed of. In this sense, man moved from direct encounter with the environment to survive, to encounter with himself as differentiated from that environment.

In its simplest terms, security needs involve those motives that are aimed at reducing the risk of uncertainty. These risks are of two types: (1) external threat and (2) self-identity. Given an incremental improvement in the capacity to consistently meet survival needs, energy is then expended to enhance security. This distinction may be seen very clearly with children. The infant is largely responsive to biological or survival demands and is wholly dependent on accessibility of the necessary external objects of satisfaction. Of course, these are mediated by other people and hence the human young are more completely dependent on others than are other species and for a much longer period of time. However, during this period, the infant does not differentiate among objects, exhibits few fears, and demonstrates very limited individualization.

After the first 2 to 4 years, the child's energies are increasingly directed at individualization and self-growth. The range of fears and anxieties increases rapidly. Competition with siblings increases for objects, attention, and affection. Frustrations as well as satisfactions are more clearly differentiated. Abstract thinking develops as does the concern with abstract concepts.

This analogy with child development is a far from exact representation of the larger human process. However, it does demonstrate the distinction between survival and security needs. Basically the two classes are similar in the sense that they define deficiencies that man seeks to eliminate. The distinction lies in how man has chosen to deal with those deficiencies.

It should be obvious that feelings of being uncertain, unsafe, or anxious emerge from a sense of disorder, unpredictability, and threat from the unknown. Projected onto the physical and social environment, these deficiencies should generate a strong motivation to construct and impose order on that environment. Furthermore, once such order is imposed, it becomes an important abstraction to be maintained and deviations from it become a threat that must be resisted and dealt with. Members of the society must be trained to perform in ways that conform to the order, and if any do not, that is the basis for punishment or rejection. For example, to regularize family relationships and reduce conflict over sex, early Hebrew culture considered adultery a major crime. Any woman guilty of this crime was taken outside the village and stoned.

Another aspect of this drive for order is the increasing value of knowledge. Given a world poorly understood, and a need to reduce uncertainty, the more knowledge and understanding of that environment that was made available, the more secure people were able to feel. Thus the value of learning and exploration increased in importance. Similarly, invention that improved the physical security of man was also of high value. Hence behaviors that were maladaptive for a survival age became highly adaptive for a security age and they were both prized and rewarded. Thus valued were not only strength, courage, and stamina, but also the more abstract "technical" virtues. The ultimate result was a markedly increased differentiation in human function and increased opportunity for individuals to find a path to self-development.

It may be noted that the skills that were rewarded were very direct ones; ones that were immediately useful to man and his society. There was little reward for abstract concerns like science or art. Occasionally tolerated, these were rarely rewarded extensively by the society or considered of great value. Invention, on the other hand, was highly rewarded and was a means of acquiring social status.

Order within the context of society, especially as it grew in size and differentiation, demanded additional sets of skills. Among these were agents of social control. Historically, these have been parental substitutes, for example, priests and royalty. These agents were almost essential to maintain continuity and adjudicate conflicts arising out of competition among members of society.

A final class of skills required in a security society was the military. Highly rigid social orders based on needs for stability and safety tend to generate highest anxieties over external threats. These anxieties generate high value for protectors against the nameless dread of outsiders. In essence, belonging and not belonging are highly differentiated constructs in a security society. The military class was the focus of the protection against outsiders. Furthermore, strong military leadership increased in-group solidarity and strengthened intragroup discipline. Royalty again was the dominant source of this military leadership. Such great value was placed on successful military action that almost every security society has rewarded it. Military actions always represented the highest measure of successful competition. It was the acting out of the basic security needs, a real and symbolic proof of the society's capacity to ensure its own safety.

These characteristics of a security society really define a world view for man's responding to uncertainty. The overriding anxieties that derive from disorder motivate a search for actions, physical and symbolic, which reduces those anxieties. Invariably these are behaviors that define what is acceptable and unacceptable—who belongs and who does not. In essence, it is control that dominates social organization in security cultures.

Although the exact formulation of the mechanics of control are idiosyncratic for each society, a highly formalized control process exists. The doctrine of the rights of kings or concepts of the infallibility of the priesthood or the Roman Senate are merely different, albeit comfortable, expressions of the needs of a society for authoritarian control. Similarly, a particular religious dogma is both a formalistic set of abstract explanations of the universe and a set of authoritative rules of social conduct. These are also mechanisms for social control whose variety is merely symptomatic of the diversity of human experience. Thus the various political forms spring from a common set of security needs and each performs the same function for their congenial social host. Conflicts between societies occur in part because security societies need to test their validity and in part because alternative forms inherently threaten each other. To be secure and safe one must be "right." To be right means singular, unique, and exclusive. It is not possible for there to be two alternative truths. The remarkable concern for and volume of discourse on political form, structure, and operations over the past 400 years of Western history merely bespeaks of the strong need for order and control. It is also indicative of the tremendous instability inherent in security cultures.

The concept of hierarchy of power and control underlies all security cultures. To a significant degree, this emerges out of the need to regularize behavior—to make human transactions predictable. This, in itself, reduces anxieties and threat from interpersonal contact. It also emerges out of a need to increase the likelihood of cooperative behavior among effective strangers. In security cultures, differentiation of function has reduced both the significance of tribal and familial ties and diminished direct knowledge of who, as well as what, one person is in relation to another. A farmer has a life style radically different from a banker and need not, and usually does not, understand banking. There is a social distance between the two members of the same society. More important, there is often a dependency relation between them. How to establish security for both in that dependency becomes an essential social concern. A man who does not raise and butcher his own meat must depend on someone who does. If their exchange is made on the basis of weight, the buyer must be satisfied that the weight is accurate. Otherwise he will have all kinds of anxieties about the transactions. Hence every "civilized" country has a bureau of weights and measures.

In sum, differentiation of function breeds a variety of dependencies among strangers, which in turn generates anxieties of varying intensity. The only way to reduce these anxieties is to have a control mechanism, a protector, a source of power that ensures conformity among people to those values and behavior that make transactions safe and secure, "fair," and "honest."

There is essentially only one way to structure a society that induces behavioral conformity. This is through operant conditioning techniques. Essentially, this is simply a process by which observed behavior is rewarded or punished in relation to its conformity with or deviance from desired behavior as defined by an external agent or accepted standard. Suppose an external agent—a Skinnerian psychologist—wishes to train a pigeon to extinguish a light by pecking at a switch plate. To accomplish this, he starves the pigeon who then instinctively pecks at things in search of food. As the bird randomly pecks the switch plate and turns out the light, he is given a pellet of food. He continues on the random pecking pattern until again by chance he pecks at the switch plate and is rewarded. Eventually, by dint of brute patience, man and bird come to an understanding. The bird rewards the man by consistently pecking at the switch plate when hungry, and the man rewards the pigeon with a dot of grain. If, however, the man stops providing the pigeon his reward, the pigeon slowly stops pecking at the switchplate. In psychological parlance, the conditioned response is extinguished.

In the human social context, the basic conditioning process is the same. Different are the variety of reinforcers that are used, the diversity of roles which involve some degree of conditioning, and the frequent use of punishment (physical or psychic pain) as a reinforcer. However, in the end, the conditioning process for man is little different from that for pigeons and is the basic method of social control.

Again, security needs are so potent after survival needs that their satisfaction dominates man's actions. He is, therefore, most sensitive and responsive to conditioning. Also, because the needs are so great, he is willing to sacrifice much of his freedom of action to obtain security. It should be noted that often this is done not consciously, but unconsciously. More important, the conscious behavior is dedicated to acquiring the security, real or symbolic, and accepting that as a good. There is normally little question of there being a cost to the process. Only poets and philosophers occasionally raise such a question. This is another reason why they are so little valued by security cultures. In general, it may be added, intellectuals are suspect in such societies because their activities are fundamentally alien to the needs of those societies to obtain fulfillment of their security deficiencies.

The conditioning process in man is carried out rather continuously. It is characterized by the method of reinforcement. Generally, three methods are hierarchical: direct, indirect, and symbolic. This first is normally used with children. Behavior is directly rewarded or punished. For a particularly desirable action, the child may be rewarded by a parent with a piece of candy. For a particularly undesirable behavior, the child may be slapped or otherwise shocked to modify his behavior.

Because most reinforcement occurs informally and usually in areas of emotional importance to a parent, the nature of the conditioning process is charged with feeling rather than being purely mechanical. The intensity or magnitude of reinforcement may bear little consistent relation to the child's overt behavior. An important consequence of this is that the particular behavior to the child is important not because of its social significance but because of its emotional significance. Furthermore, since children especially have great difficulty in differentiating objective consequences of behavior from the immediate emotional response of the conditioning agent, they often fail to associate real behavior with real consequences. As one result, the process decreases security.

Reinforcement is also inconsistent. Children are generally conditioned on a partial reinforcement schedule. This occurs both because parents are involved in other than supervising activities and because their own emotional state varies. Hence the impact and significance of any one instance of behavior may generate a highly unpredictable magnitude of reinforcement. Partial reinforcement can lead to very different conditioning results. Above a certain rate, it can produce a very permanent or long enduring conditioned responses. Below a certain rate, depending on the behavior, it can produce, not learning, but maladaptive emotional response.

Because of man's exquisite capacity to communicate both thought and emotion through language as well as physical action, that device is the most frequent means of reinforcement. Verbal praise and verbal punishment, for example, threats, are very effective conditioning stimuli. They are also very economical. In most cases, it is both easier, quicker, and more natural to use speech than physical objects as rewards or punishments. Furthermore, linguistic conditioning can more easily be generalized.

Indirect conditioning occurs through peer groups and usually by example. In any social environment, individual members are constantly observing both behavior of others and group responses to those behaviors. In cohesive or close groups, the approval or rejection of those responses is felt and acts as a secondary reinforcer. In social settings, it is these indirect participatory experiences that produce social learning. In most societies, many of these occur within formal institutions, mainly work or schools. The latter are, to a greater or lesser extent, devices for providing children controlled environments in which social values may be learned. However, most indirect learning goes on within informal groups. There is considerable evidence to indicate, for example, that among adolescents, criminal behavior is a product neither of ignorance of "right" nor psychological maladjustment. Rather, peer group rewards and punishment develop independently of larger cultural values. Behavior that rewards the group member may coincide with the larger social group values. When it does not, the group or the individual isolated from it is considered deviant or criminal. The young individual is called a delinquent and the group labeled a "gang." Where the group shares the larger cultural values, it receives the reward of support and approval from society. Yet, in all cases, the need for security motivates individuals to sustain the rewards from the peer group even when it contradicts larger social values. That is the measure of the power of the peer group.

Given the structure of the learning process, it is not surprising that eventually institutions are imbued with the power to provide rewards and punishments. The institutions generally use symbols rather than actual or direct reinforcement. Almost by definition an institution is created for that purpose. A society of engineers has power over its members both internally, by setting standards for admission, and externally, by representing a body of expertise which speaks with a single voice equal at least to the sum of the individuals who make it up. An individual engineer has little intrinsic authority, but an association of many engineers does. There is no rational reason for this, but in a security culture, a formal group does have such power.

Almost all security societies are organized on a hierarchy of power, that is, the symbolic capacity to provide rewards and/or punishments. It is position that determines this power. The man who is in that position has the power because of it. It is a fundamental organizing principle in Western society, at least, and appears characteristic of security cultures in general. The society is organized in a pyramidal fashion in which power is increasingly attributed to fewer and fewer people. It is this fact that provides learning by symbolic position.

In security cultures, because of the symbolic capacities of position, there is usually great ambivalence toward power symbols. On the one hand, they are desirable, because an individual who can attain such a position increases his own security. On the other hand, an incumbent in a position must exercise power and hence constrains the behavior of those "below." That individual will thus be a threat to his subordinates. He can command and the subordinate must follow.

A teacher of middle-class American youth has remarkable power over those children. He or she can determine what the student may do; he defines what the students must commit to memory; and he defines success or failure for them. All of this is done symbolically by such things as gold stars or grades. (It may be noted that this process rarely works with children having different basic values, such as those brought up in a ghetto culture.) Also the hierarchy in the federal government rather rigidly prescribes the maximum size of office space by rank in that heirarchy. The presidents of major universities almost always have chaffeur-driven automobiles, but deans of the various schools in those universities usually do not.

On balance, in security societies, power hierarchies are acceptable so long as the constraint exercised over members' behavior is within a tolerable range and when the exercise of that power is consistent with the social rules for its extent. For example, someone who steals a loaf of bread may be punished by say 30 days in jail, but if a judge were to sentence the man to death, his exercise of power would not be socially acceptable. Every security culture sets implicit limits on power and generally tolerates excesses only for relatively short times. Within these limits, security is enhanced and members can predict consequences of behavior—theirs and others—more successfully. Since the goal is personal safety and orderly growth, such hierarchical control mechanisms are acceptable and generally essential.

The basic operant conditioning process that underlies behavioral control in security cultures leads to a reward reinforcement basis for behavior. People act in behalf of others because it is "right" or because it is approved. The behavior is rewarded and the expectation is that the reward is direct and immediate. Conversely, people refrain from certain behavior because of fear of punishment. One does not steal because one might get caught. An individual should go to college because the college graduate earns a quarter of a million dollars more over a lifetime than does a high school graduate.

In security cultures, the emphasis on direct reward and punishment places a high value on exploitation. It is a proving and acquisitive frame of reference. The security deficiency motivates competition and self-testing. And it places great value on possession. Thus it is quite acceptable to exploit the natural environment to enhance the security of the group. Similarly, it is quite acceptable to exploit people and groups who are aliens to enhance the security of the society. Indeed, those members who provide the means to accomplish such exploitation are highly rewarded.

Out of this total process, these societies emerge as scalar entities. That is, they order all behavior on scales of worth. Every form of social behavior is categorized and ranked from positive to negative, good to bad. These scales form the matrix by which member and nonmember may be judged, and they provide the essential basis for defining the level of security of that culture.

The crucial attribute of security societies is their pervasive property of evaluation of behavior. The history of Western society is the history of the development of (1) alternative means of defining the dimensions to be evaluated and (2) the social organization for imposing those dimensions on the mass, in actuality as well as symbolically.

Such a process can proceed, if and only if, there are acceptable and invariant principles on which the evaluations may be based. There must be a truth-set agreed on explicitly or implicitly by which all actions may be judged and all behavior controlled. Once this is accomplished then a means exists for inducing conformity of behavior. It should be understood that such a process has been highly adaptive in differentiated societies. It provides order, defines the basis by which all receive reward and punishment, and permits coalescence of people into cooperative groups. It is, in principle, a design for a social amplifier, allowing diversity of function to be aggregated into coherent function and toward common ends. The end, of course, is increased safety and security.

Just as soon as this organizing principle was discovered, the size of social groups and the number of their functions increased rapidly. The early urbanized cultures in the Eastern Mediterranean had total populations under 500,000. Athens had a population at its height of under 50,000 citizens. (It also contained over 300,000 humans who were slaves and aliens. Of course, they were counted only as domesticated animals.) It was, in fact, in the city-state of Athens that the diverse lines of evaluative principles were synthesized and implemented. Greece is less important for its notions of democracy than it is for creating the formal basis of rationalism.

Clearly, what man does to satisfy his security needs, especially to other men, must be formally rationalized. In the West, it was Greek thought that was the basis for this rationalization. The formalities of expression and means of implementation have been expanded and embellished continuously over the past 3000 years, but only the efficiency has gone up. Western security society has progressed from the slave state of Greece, to Roman law, to the Holy Roman Empire, to the imperial states of Western Europe, to the economic technocracies of present time. All are evolutionary steps in the formalization in rationalistic terms of man's continuing search for safety and security. The only substantive change has been that man's demand for security has been resolved through immediate gratification and continuity of reward, whereas 500 years ago man had to console himself with heavenly rewards. Consequently, engineers and scientists, by creating the means of providing both increased security and safety, have themselves been rewarded with higher status, monetary value near the top of the professions, and, until the midtwentieth century, high trust among the membership of the society at large.

Self-Growth Needs

What has been discussed up to this point is a concept of human motivation largely based on the concept of deficiency needs. The basic biological needs and the needs for security and ego realization underlie man's behavior and how he organizes his society. However, there is another level of needs that are radically different from the deficiency ones. These appear to involve concepts of the individual as an independent agent, seeking a sense of self. There seems to be a need within individuals for psychic growth. There seems to be a need to experience, to feel. It is a curious property of this class of needs that evaluation and rationalization disappear or become irrelevant. Individuals operating at this level are self-motivated, quite insensitive to traditional social and physical rewards. Furthermore, their interpersonal relations appear to involve not transactions but interactions. It is self-growth rather than deficiencies that must be filled.

It would appear that when, for whatever reasons and by whatever means, man's deficiency needs are satisfied—in the stochastic sense at least—he ceases to want in the specific sense of objects or states, but rather needs to realize and express himself.

When this level in the hierarchy of needs is reached, man's values markedly change. What motivated him in the security level no longer matters: success, status, attention, affection, power, possession, rightness, approval. The individual in school is not there to become an engineer or an economist; to find his path to success in the world; to develop skills to serve mankind; or to prove himself as good as or better than anyone else. Rather, he seems to be there out of curiosity, interest, and exploration. Individuals operating under self-growth motivation tend to be less concerned with achievement, consistency, or perseverance. In the educational process they tend to be unconcerned with grades, to drop in and out of school, and to engage in multi- or cross-disciplinary studies.

There is increasing evidence that creativity and level of growth motivation are highly correlated. So also are original thinking and concern about relationships among concepts. People operating at this level of the need hierarchy appear able to generate more solutions to problems and use more resources in defining solutions than those operating at the security need level.

Because of sharpened awareness of relations in particular, people operating at the self-realizing level are more conscious of both the short-term and long-term consequence, of actions. The concern about ecology, and citizen actions in its behalf, did not arise until the midtwentieth because security-need deficiency precluded it. Jobs, income, production, and profit were too central a concern for there to be any critical mass of public activity over air or water pollution. Only if such pollution were a direct threat to life or property would social action be initiated. Indeed, the first major air pollution control programs occurred in London only after a major episode caused the death of several hundred people.

People operating in a self-actualizing mode would need no such justification to act to obtain improved environmental quality. They would act out of the direct awareness of a negative or destructive interaction between any of man's works and the environment. There does not need to be a rational justification to protect any part of the natural environment. Only a security culture carries out a benefit-cost analysis. A self-realizing culture rejects immediately alternative decisions that impose a cost greater than zero, regardless of the benefits. If you will, a self-realizing society refuses to accept the lesser of evils in social decision-making or in engineering works.

This level of sensitivity extends as well to human relations. Self-realizing people are motivated in their relations with others, not out of sympathy, but out of empathy. Empathy is the capacity to feel what another is feeling, and action toward people emerges out of that sharing of feeling rather than out of need for or expectation of reward.

It is probably because the Western world is emerging into a self-realizing culture that so many traditional cultural norms are changing. If people accept idiosyncracy as a positive attribute, if people highly value individual identity, and if people can empathize with others, then acceptance of such concepts as abortion, sexual freedom, and individual choice of life style, and rejection of regimentation, coercion, conformity, and social control are to be expected. One no longer finds it acceptable that decisions be made merely for administrative convenience or on traditional economic grounds.

The social organization that emerges out of a self-realizing need structure is radically different from those extant in Western history in the past. One need read such modern writers as diverse as Marcuse, Reich, Maslow, May, Buber, Tillich, or Boulding to understand some of the implications. For such a society will be nonlegal, noneconomic, apolitical, self-responsible, permissive, and experiential.

As the Western world moves toward a self-realizing culture or a "post civilized" society, the role of politics-and-technology, which is the controlling force in the West, must shift and is shifting. The problem for this dominant partnership, whose roots are in the logic of the Greece of antiquity, is how to cope with increasing diversity and to rationalize incommensurables. At the minimum, a broader and more comprehensive logic for social decision-making and a more sensitive and adaptive technological process are required than that in use up to the midtwentieth century.

Summary

In sum, the purpose of this discussion has been to provide in a most superficial way an alternative view of human behavior that may improve the perspective on the emergence of human society. It is a motivational frame of reference which emphasizes the basic need structure. Social evolution is a consequence of the experience of man in developing means to satisfy his needs. We see, in effect, three levels of needs: survival, security, and self-realization or growth. Each level leads to a different social organization and a different set of common values. When we talk about societies operating at different levels of the need heirarchy, this should not be interpreted that each level is discrete or uniform.

It should also be recognized that where man may be in the need hierarchy, either as an individual or as a whole society, is reversible. Man may be reduced to subsistence level as well as be raised to a security or a growth level. These changes occur because of physical, social, or psychic events. The overall direction of human history does appear to be, however, from the lower to the higher level.

Finally, the transition from one level of the need hierarchy to another is not discrete. Rather it is a continuous and emergent process. One reason there is often great conflict during periods of transition is simply because new values and priorities are so different from the existing ones. There are then, by definition, at least two cultures threatening each other. When this happens within a society, there is not only debate, but also polarization, which slowly leads to decreasing trust. The social ground rules by which men carry out their transactions become suspect. Consequently, the society is required to expend much of its energy maintaining the integrity of the basic cohesiveness of the group. If the methods used in the face of irreversible change are largely repressive, the probability of rebellion is increased. If the society, in the statistical sense, is secure enough, adaptive responses will occur that seek some level of compromise and accommodation. This process leads to revolution, that is, a change in cultural values and a restructuring of social organizations or priorities.

SOCIAL ORGANIZATION

It should be obvious from this discussion that to increase the stability of man's existence and to fulfill his basic needs, he organizes his social as well as physical environment. In survival societies this organization is simple and directed at two very simple ends. One is to divide labor to attain sufficient survival materials. The second is to provide for common defense against external attack.

In security societies, the same two functions are required. However, there is a marked increase in the diversity of functions required. The range of interpersonal relations increases in both numbers and complexity. In addition, security societies require a social organization that maximizes order and predictability. It does then place increasing emphasis on behavior control and on training or education.

In growth societies, the same needs inhere as in survival and security societies. However, order and predictability are of an implicit form. It is assumed in current conceptions that man is basically orderly and his behavior is directed toward his own growth. Hence his behavior is always positive, constructive, and creative. The issue within such a society is to maximize the likelihood that the individual members may attain their psychic growth. Since it is an empathetic society, the members are supportive and sustain each other. Because there is mutual awareness and understanding of each other's deficiency needs as well as growth needs, there is little competition and little threat, since no one takes advantage of another. Consequently, growth societies are inclusive ones rather then exclusive, as security societies almost of necessity must be.

Organization of societies implies some form of governance. It is a basic function that inheres in all social orders. However, government plays a variety of roles which also depend on where in the need hierarchy a society is. Within that context, we can identify four historical functions of government quite independently of its form or political philosophy.

The first has been mentioned on several occasions earlier in the discussion. This is government as the focus of social values. In essence, it is a ruling system in which the leader is endowed with power and control which he can exert over the polity very directly. In this case, government is a master demanding allegiance and specifying acceptable behavior. It is generally acceptable because it symbolically enhances security and practically ensures cooperative behavior. Leadership in this sense is usually charismatic; the individual is endowed with personality so unique that it arouses attraction, awe, and following behavior.

The second function of government is as an agent for adjudicating intragroup conflict. In survival and low order security societies, social organization and communication among members are unstable and ambiguous. For a variety of largely psychic reasons, it is impossible for members to resolve conflicts which arise among them. In these societies, truth and justice are on polar scales which allow little middle ground. Hence members become trapped in irreconcilable conflicts. In cases of serious conflicts, an external agent must intervene to resolve them. The agent usually cannot be an innocent bystander, for such would have no symbolic value to the contestants. There must be an agent who has power to resolve the underlying psychic polarity. This agent must be a source which has special power to divine a truth that adversaries, that is, normal people, are incapable of. Government is uniquely set up to do this. It may be formally set up as by a king, priest, or his carefully designated agent or by an explicit codified rational process established for the purpose. Systems of laws have emerged, especially in security societies, for just this purpose. Again, mechanisms for formal codes of laws reduce uncertainties about judgment and allow people without charismatic power to maintain order efficiently.

A third function of government is to determine and set priorities for the society. It is the focus for identifying social priorities. At its best, this involves a process of communication from the constituents of the society through which the governors obtain information. They then may evaluate that information and articulate the priority needs. The interesting aspect of early Greek democracy was the structure by which citizens were given a "voice." It was a communication network among a very small group that could be used by the leadership.

At worst, priorities are set in government by individuals with little or limited knowledge of the realities of social life. Priorities are set on the basis of the values and perceptions of the leadership. Depending on who the leader is the leadership could be intuitively sensitive to social problems, needs, and priorities, or it could be largely antithetic to them. In most historical cases, the distinction between beloved king and hated tyrant has been simply this. Government reflects the values and priorities of the majority of the society or it cannot govern; it can at best only control.

In either eventuality the governing force by word and action is responsible for articulating these priorities. It is government's responsibility to set the goals and to obtain consensus for them. If the government cannot do so, it will eventually be replaced.

The fourth function of government derives from the third. It provides the vehicle for implementing the goals of the society and providing the means of satisfying the members' security needs. Regardless of how resources are obtained, government must allocate them to actions or programs according to defined priorities. At its simplest, this allocation is done directly, by forcing people to perform certain functions, or by law or edict, which defines how people shall act to obtain the desired goal. Laws, because of their implied punitive power, are a most common means for directing or controlling social energies to obtain social goals. In the United States, society, because of its legalistic roots, has used this technique explicitly to obtain conformity of behavior, to solve conflict problems, and to institutionalize values. For example, because of the central significance of family to the continuity of the society, society has made marriage a very simple legal act while divorce has been made very difficult.

At its most complex, government uses resources to establish implementing agencies which then apply specific techniques to the solution of a priority problem. In essence, the government is functionally organized to implement programs of supposed benefit to the public funded in some relation to immediacy of need and social priorities. The mechanisms for meeting priorities so institutionalized are usually traditional ones that have been proven useful in past experience. Only in a few cases have resources been allocated to the development and implementation of innovative solutions. Only in the national security effort, and to a lesser degree in health and agriculture, have the priorities been met by innovative programs.

THE TRANSITION PERIOD

It should be clear from this look at the course of the development of man and his society that in the midtwentieth century, we were at a point (in the United States, at least) of major transition. There was increasing evidence that the society was emerging out of a security need structure toward a growth society. The major and rapid changes in values toward greater permissiveness and individual privacy are obvious examples. So also was the growing sensitivity to what hitherto were marginal concerns. Awareness of and investment in education, health care, and the problems of the elderly and handicapped; concern for the economically disadvantaged; and the protection of the physical environment are all examples of increasing empathy. Almost none of these priority issues really has a direct reward, economic or physical. In sum, priorities are changed or changing because the society was in process of transit to a new level in the need hierarchy of man. We were moving from a security to a community society.

A third implication is that technology utilization for productivity and/or profit alone is being seriously attacked. The whole meaning of the movement pioneered by Ralph Nader is that human safety, dignity, and individuality can no longer be sacrificed by physical or administrative engineering for abstract, exploitative goals. Public policy decisions that abet this are also attacked. The growing rejection of an inviolate highway trust fund as a special interest and a too narrow means for meeting mobility needs is an obvious example of this change.

A final implication is that public policy and goals predicated on aggregate statistics or averages are also being rejected. The treatment of man as a class or as an object is generating considerable hostility. The rejection of schemes of human classifications which were set up for administrative convenience simply shatters a value that has endured for a millennium. System design must now be based on satisfying individual needs. Similarly, public policy decisions increasingly have to be responsive to individual objectives and goals.

These changes in social goals and values impose profound burdens on public policy. When the goals and priorities of society were simple and based on physical security, the number and complexity of both the decisions and the allocation process were small. Intuitive, superficial, and short-term decision-making was sufficient. Traditional consensual processes such as legislative acts derived from testimony or individual insight and perseverance were sufficient. To direct a mass required only an approximation of its center of gravity.

A far more rational and explicit policy making procedure is as essential as redefinition of leadership. It would appear self-evident that intuitive analysis of priorities and policies cannot provide adaptive solutions to highly complex social problems. It can only exacerbate them. For example, the United States government has pursued a policy of public housing that was based on simplistic judgments without even a superficial knowledge of the social forces at work, and as a result the policy has compounded the disorganization of disadvantaged social groups, increased crime, and built social traps rather than springboards to growth. The public housing policy of the midtwentieth century became recognized as one of the greatest failures of public policy and resource investments of the century. In essence, public policy must be made on far more knowledge and on far more rational analysis procedures than has been done in past eras.

One of the first things that needs to be done is to separate what people think a social problem is from what in actuality the problem may be. Five hundred years of history in the physical and behavioral sciences has demonstrated that testimony or consensus judgment alone does not determine truth. We have used these techniques for nearly 50 years in traffic safety and generated the myth that engineering, enforcement, and education could eliminate accidents. We have built a monumental administrative structure whose overall production in terms of accident reduction has been negligible. Without a commitment to systematic analysis and understanding of underlying causes of phenomena, no rational policy-making is possible. The results are a waste of limited resources on wrong or ineffectual solutions. Or, at least, inadequate policy-making leads to investments having adverse external consequences of significant long-term impact. Water pollution is the classic case in point. If in the process of solving one problem we create one or more worse problems we have made bad policy decisions. Such has been more the rule than the exception in the Western world in this century.

Thus a highly analytic and comprehensive policy process is an ultimate requirement if new values and new priorities are to be defined and met. In this process, science and technology will have to function in a quite new way. But to bring about this change a new kind of partnership between technology utilization and public policy is required.

SELECTED READINGS

7.1 Banfield, E. C., *The Unheavenly City,* Boston: Little, Brown, 1970.

7.2 Bronwell, A. B. (Ed.), *Science and Technology of the Future,* New York: Wiley-Interscience, 1970.

7.3 Buber, M. *Between Man and Man,* New York: Macmillan, 1955.

7.4 Cox, H., *The Secular City,* revised edition, New York: Macmillan, 1966.

7.5 Fletcher, J., *Situation Ethics,* Westminster Press, 1966.

7.6 Lindsey, G. Y., and E. Aronson, (Eds.) *The Handbook of Social Psychology,* New York: Addison, 1970.

7.7 Marcuse, H., *An Essay on Liberation,* Boston: Beacon, 1969.

7.8 Maslow, A. H., *Toward a Psychology of Being,* New York: Van Nostrand, 1962.

7.9 Nobile, P. (Ed.), *The Con III Controversy,* Pocket Books, 1971.

7.10 Randall, J. H., Jr., and J. Buchler, *Philosophy—An Introduction,* New York: Barnes and Noble, 1942.

7.11 Reich, C. A., *The Greening of America,* New York: Random House, 1970.

7.12 Schooler, D., Jr., *Science, Scientists and Public Policy,* New York: The Free Press, 1971.

7.13 Skinner, B. F., *Beyond Freedom and Dignity,* New York: Bantam/Vintage, 1971.

7.14 Toffler, A., *Future Shock,* New York: Bantam Books, 1971.

CHAPTER 8

POLICY DEVELOPMENT AND THE POLITICAL PROCESS

Politics and political systems provide the decision role in the difficult area of choosing between alternate paths that a group or society follows. Because of limited resources and capabilities, and because of differences in the opinions of individuals as to what is "good" or "bad" (or important and unimportant), political judgements must be made.

A very simple analogy to the political choice problem is the decision individuals make in choosing one item from a basket of mixed fruits. Which to choose, or whether to take an item at all, is a personal judgment that embodies the problems of saving for the future, consideration of health, and individual preferences. What one chooses is inseparably related to one's preferences, interests at the moment, and attitude toward the needs and desires of others. There is no single right or wrong choice.

The problem of choosing between alternative paths exists for individuals, families, and groups of individuals. The question of how to resolve public conflict or differences of public opinions is typically viewed as a political problem. The power structure or controlling influences emerge, governments are formed, and associations and corporations arise. These institutions, formed for a variety of reasons, have internal political systems, and collectively provide the means for individuals and society to advance or degrade.

There is no way to avoid the difficult problem of choosing, so there is no way to avoid conflicts and compromises between individuals and groups. In this chapter, the relation between technology utilization and the political process is examined, and the principles that control decision-making as related to the provision of engineering works are considered.

POLITICAL SYSTEMS

The political process consists of principles, procedures, and techniques that are used to control or govern a group of individuals. As used herein, political processes do not refer to polity or political systems. Political systems are organizations using sets of rules or practices developed for the purpose of governing. These systems are controlled by the political philosophy or convictions under which a society chooses to operate.

Political versus Technical Decisions

Regardless of the political system, the political process consists of the following phases:

- Evaluation of existing conditions
- Assignment of priorities for improvement
- Identification of alternative means
- Allocation of resources
- Establishment of laws, rules, or regulations
- Evaluation of the effectiveness of the action

If a problem can be solved analytically, it is not a problem that involves the political process. The political process is a technique for developing solutions to problems that require predictions of the performance of social as opposed to physical systems.

Decisions based on the political process require political judgment which can be defined as a judgment as to which of two or more alternatives is in the public interest. Since both action and inaction are alternatives, individuals, groups, and governments repeatedly make "political judgment" as defined herein.

The common characteristic of these judgments is that a choice must be made on quality of life, and no rule or theory exists for a general solution to the problem. If a political judgment (or assumption) is made, then there is a basis on which rational analyses can be made. For example, in industry, a political judgment can be made as to the margin of profit that a company is to make. This judgment cannot be explicitly set, and is influenced by a wide range of social, economic, environmental, and political considerations. After the judgment has been made, analytic techniques can show the alternative ways to achieve the objective. Similarly, governments can make political judgment as to the desired purity of the environment versus other benefits, and analytic techniques can provide alternative means to these ends. Government problems in particular are complex, because no two alternatives can be perfectly equal, and therefore, for precise objectives, at least two levels of political judgments are required—general and specific.

As noted earlier, the political process includes predictions of performance of social systems and ultimately a decision that influences the future performance of those systems. Technical processes also include predictions of the performance of systems, and decisions that influence future performance are frequently included in the technical process.

An important distinction should be made, however, between technical and political decisions. This arises from the fact that a technical decision constitutes an allocation of resources, which in turn influences the availability of resources for other purposes. Accordingly, *a purely technical decision cannot be made until someone, or some group has made a political decision as to the ground rules by which technical decisions will be made.* The following important principles, therefore, can be stated:

1. There is no technological basis for deciding how or if resources should be used.
2. Decisions that lead to the use of resources are political decisions.

3. A purely technical decision is one that:

   a. Does not involve an allocation of resources, or
   b. Is based on ground rules established by a political decision

4. A political system's delegation of authority to make decisions that allocate resources does not change the fact that the decision remains a political and not a technical decision.

5. If technologists are to directly influence the course of civilization, they must influence (or make) allocations of resources. Therefore, to the extent that an individual is dedicated to working in the public interest, he must become a part of the political process and existing political systems.

The Nature of Political Power

In all political systems—large or small, in government or the private sector—there is an ultimate source of power or decision-making mechanism. In some systems, this can be a group of individuals, but in large organizations, one individual must approve or disapprove a course of action. Leading to these decisions there can be a variety of forces or even a single individual that virtually controls the decision-maker. Thus the power structure can be both complex and obscure, but is a necessary part of a political system that governs a group.

The existence of a power structure and its ability to control those who are governed are two different issues. Certainly a power exists, but also, individuals under even the harshest tyrant will defy the law. Therefore, we are talking here about decisions that *are* made, and the purposes or objectives of those decisions.

Two types of tools are available to a political power for controlling the progress of a society:

- Control of the pool of resources
- Laws and regulations

Debate and rhetoric are useful techniques for persuading individuals to follow a course of action, but are primarily useful for retaining power or directing uncontrolled private resources toward a desired activity.

The control of resources and the setting of constraints (laws and regulations) can be potent weapons. In a given community, the power structure may well be split between government and the private sector. With regulations, the government can partially control the power of the private sector, but the balance of power will generally rest with those that control the most resources.

The relation of technology to political power is the significant issue for the purpose of this discussion. No matter what form, no matter how complex, no matter how diffused, technology is ultimately the tyrant who *is* the power or the servant of the power that controls resources. For technology to be directed to a specific purpose, the power structure need to so determine and state.

A dedication of the technologists to the public interest requires a knowledge of the political power with which he must work. The following principles can be stated:

1. Unless the scientist or engineer is the duly authorized or a part of the political power structure, there is no basis upon which he can either assume the responsibility for or be held accountable for the results of the process for allocating or regulating the use of resources.
2. Insofar as his area of competency is concerned a scientist or engineer should provide the decision-makers with as complete an analysis as is possible of the consequences of a proposed use of resources.
3. Approaching the decision-makers in an advocacy role relative to a given use of resources is a political not a technical activity, and is ultimately governed by political processes.
4. To the extent that an advocacy role from a scientist or engineer prevents a decision-maker from obtaining a complete analysis of the consequences of resource allocation, questions of morality are based on political, not technical, definitions of integrity.

Goals, Resources, and Constraints

Goals, resources, and constraints form an inseparable set of conditions which are an implicit or explicit part of the policy guidelines that control a course of action. If two of the three are specified, the variability in the third is quite narrow. If only one is specified, an infinite number of alternatives exist. To achieve the greatest effectiveness, the political process needs to provide quantitative limits for each of the three. This principle is discussed further in Chapter 9.

GOALS

At each level of political power and within each subsystem of a senior system, the goals must be established. For the highest levels of government, the total social system is involved, and the goals need to be understood in relation to the quality of life of individuals, groups of individuals, and the entire community. Thus the mayor of a city may be concerned about poverty, public safety, health, education, and so on. Furthermore, he is concerned about specific conditions relative to some characteristics or groups in terms that reflect the well-being (or the hardships) of individuals.

Given alternative goals that can be considered, the alternative means (or subsystems) to achieve the goals become relevant. Thus goals for transportation, communication, energy, and such may well be desired to achieve broad social goals but are means to some other ends, rather than goals in themselves. Likewise, objectives for hospitals, police, and school buildings are derived from such goals as increasing life expectancy, reducing the amount of crime, and increasing the level of education of all or segments of the community.

A description of a subsystem is that it constitutes one means (part or component) for achieving some end (objective of a senior system or larger subsystem). Certainly subsystem goals should be defined, limited, and scheduled in relation to their significance in achieving the goals of a senior system. Furthermore, a change in subsystem performance should be fully understood before efforts are made to estimate the effect on the performance of a senior system. Since more than one subsystem exists in most instances, more than one alternative exists for improving the performance of a senior system.

The preceding principle can be understood by considering the transportation subsystem.

A change in the quantity or quality of transportation may be traceable to a need to further certain social system goals. In such a case, the change in quality of service (output) of the transportation system should be considered first, rather than the number of buses or miles (or lanes) of roads or streets. It is a change in income levels, public safety, health, and so on that generates the social goal, and this comes about through (1) faster service, (2) more reliable service, (3) less costly service, and such. Many alternative transport solutions could provide the change that fulfills the social systems goal. Given only the change in output that is desired, a much wider range of alternatives is available to the decision-maker.

Perhaps no other principles of goal-setting are more critical than the following two factors. *First, the senior system must control the goals of its subsystem. Second, overcontrol of a subsystem (specifying more than a change in output) can result in seriously reducing the effectiveness of the senior system.* But in a complex society, for example, the federal government, the setting of social goals in terms of changing expectancies of individual well-being can be quite difficult. The problem is compounded by self-interests of subsystem hierarchies that are aware only of the improvements needed in their own subsystem, and provide strong recommendations for resources to make the changes they desire. This practice can lead to a failure to focus on the ills of society and the needs and desires of individuals.

The same problem exists at all levels of government. Community goals are frequently viewed as subsystem improvements, and the ultimate benefits that are desired are obscured by the search for benefits from subsystem improvements.

The importance of setting goals is reasonably well accepted. The significance of dealing with the output of the system or subsystems, however, is a more difficult concept to grasp. By not considering subsystem output, the subsequent encroachment and the required assumptions can result in a less effective solution.

RESOURCES

Closely allied to the goals that are set are the resources that are needed to achieve those goals. The goals-resources problem is normally solved through an iterative process; that is, the desired change is considered and an estimate prepared for the level of resources required to make such a change. The alternative uses of resources can then be considered, which may lead to a change in the goal and a corresponding change in resource requirements.

The problem can also be approached by assigning resources and obtaining estimates of the range of change in system (or subsystem) output that could be obtained. Here again, an iterative process is useful.

For the political process, estimates of the resource requirements do not have to be precise. In fact, policy decisions are frequently made with limited knowledge of the entire spectrum of technical possibilities. Particularly where innovative solutions are to be considered, the estimates of resources are necessarily approximations.

Estimates of cost of public works have been notoriously underestimated, at least as reflected by comparisons of the initial policy approval and the final costs. These underestimates can occur for a number of reasons, but within the political process the overrun in expenditures can usually be traced to (1) a deliberate political strategy to obtain policy approval, or (2) a change in goals before the program is completed. On occasion, there have been failures to estimate accurately, including the problem of cost escalations with time.

Estimates of resources is a technical activity and need to be developed by a technologist who is familiar with the problems of modifying the performance of a particular system.

For policy analyses, the inclusion of political strategy in the preparation of resource estimates can lead to ineffective programs. Certainly a deliberate underestimate of resources for the purpose of obtaining program approval is a political activity, not a technical one.

Where innovative actions are encouraged, estimates of required resources are frequently based on the requirements to achieve the goal with little risk of failure. Alternatives that include research, development, testing, and evaluation can then be included for policy consideration, with an appropriate adjustment in resource requirements.

In a final analysis, the policy decision hinges on achieving the best combination of goals with the available resources. This step is discussed in Chapter 2.

CONSTRAINTS

Problems of esthetics, pollution, and safety have been a part of the technological process, but have also been a necessary consideration of policy-makers. In recent years, and with the advent of life enhancement (community culture) as a more vital factor, such matters have increased in importance. As a result, certain alternatives have been excluded because of the adverse effect or impacts that would result.

To complete the policy process, the limits under which a subsystem must operate should be stated, which establishes further limits on the goals-resources problem. Constraints can deal with physical environment or with any impact on the social system that is considered intolerable; for example, neighborhood disruption, intrusion into parks and recreation areas, and safety are types of effects that can be controlled.

In effect, the imposition of a constraint is a resource allocation, or if not, it results in a delaying action for the improvement of some subsystem. Constraints on subsystem are necessary, however, to control adverse social impacts. Normally, these constraints mean that an innovative solution results in increased use of resources, or reduced performance of the subsystem (in terms of its purposes), or more elapsed time (in order to include research and development). The variables of performance, resources, and time are controlled by the political process because of their ultimate effect on the rate of change in life qualities.

Since placing of constraints produces (1) a reduction in the goals or (2) an increase in resources (or both) for otherwise similar subsystem outputs, it is critically important that the constraints be set during the political process. If the limits have not been set, the range of alternatives that are studied for a given goal-resource combination will include unacceptable solutions. There is no avoiding, only delaying of, the analyses of the political consequences of applying or not applying constraints, which if included with the policy decision result in significant savings in analysis costs and implementation time.

PRINCIPLES

The following principles relate goals, resources, and constraints to the political process:

1. The goals, resources, and constraints are the major components of policy requirements for engineering works.
2. For the most effective actions, policy decisions should include quantitative specifications for each of the three components; that is, goals, resources, and constraints.
3. The policy requirements for a subsystem are not specified by the subsystem, but by policy-makers of a senior subsystem or systems.
4. Internal improvements in a subsystem should be the responsibility of that subsystem if the input and adverse effects of the output remain constant or are reduced.
5. A more effective solution can be developed, if goals are stated in terms of changes in performance, rather than specific modications of a subsystem.
6. Estimating resource requirements is a technical problem and requires technologists who are familiar with the system or subsystems to be modified.
7. Requests for resources are a part of the political, not the technical, process and are controlled by political strategy considerations.
8. Constraints are one form of resource allocation and the time or resources required to provide for the constraints should be established.
9. The balance that is set between goals, resources, and constraints is achieved by an iterative process and is ultimately a part of the resource allocation process.

Political-Technical Relationship

Because resources must be allocated, there is a major interaction between the political and technical processes. The primary concern of the political process is the impact of subsystems on the well-being of mankind. This concern reduces itself to a question of how to change subsystem effects (output) so as to achieve the most favorable impacts. Insofar as achieving social goals, the technique by which the output is changed is less important than that it be changed. The *how* eventually becomes important in political strategy, since some activities are more readily accepted than others. Nonetheless, the focus of the political process should be on the *change in output* rather than on a change in hardware. Accordingly, the political process needs to interact with the technical process to determine alternative impacts from alternative changes in subsystem outputs. With this information, decisions can be made that produce impacts that are most important to the community, system and subsystem involved.

The following principles can be stated concerning political-technical interaction processes:

1. Interaction between the political and technical processes is inevitable.
2. The decision (or opinion) about which of two social benefits should be sought is a political decision (or opinion) no matter who makes the decision.
3. The judgment as to how a change in a subsystem will influence the performance of that subsystem is a technical judgment, no matter who makes the judgment.
4. Those who advocate or argue for the use of resources for a specific purpose or who advocate a set of regulations or constraints are engaged in political activity, and to be effective will need to adhere to the ground rules of the political system.
5. Persons in the political process who make technical judgments without having the necessary or appropriate qualifications should be challenged on technical grounds to review their decisions in light of competent technical judgments.

The technical process can provide estimates of how and how much subsystem output can be changed as well as the impact of these changes. These estimates do *not* include choosing between alternative benefits. Rather, technical judgment extends only to predicting the performance after changes are made in systems or sybsystems, unless (1) political guidelines have been provided, or (2) political assumptions are made.

Correspondingly, there is no political basis for choosing between two solutions if estimates of the difference in performance between the subsystems are not available. Political judgment does not extend to the area of predicting subsystem performance. If the political process includes such a procedure, the judgment is still technical regardless of who makes the estimate. For example, a policy decision that new high-speed rail service, or new rapid rail transit, or new exclusive bus lanes should be provided, requires that someone must make the following technical judgments: (1) that the change in transport service (or an environmental impact) is superior to what exists or is likely to develop; (2) that any change in transport service is relevant to achieving social system goals. While political judgment is essential for choosing between alternative social benefits, it is a poor and unnecessary way to obtain estimates of technical performance.

Just as engineers extend their activity into the political process, persons in the political process extend themselves into the technical arena. The only satisfactory excuse for it is the unavailability of technical skills. For example, private individuals may make technical judgments concerning the construction plans for a residence because hiring an engineer costs too much. Or a county judge may be forced to make technical judgments about roads because there are no engineers in the area and there are insufficient funds to provide an engineer. If engineers produce the type of analyses required by the political process, there would be no need to extend political judgment into a problem that requires technical judgment.

Technologists and the Political System

To assure more effective decision-making, it is necessary to provide the controlling political system with the total consequences of a given course of action. But political systems are composed of a wide range of individual human beings with a variety of skills, variable motivation and loyalties, and no rigid set of theories and analytic processes whereby answers can be generated. Furthermore, the ethics of the political system is not necessarily the same as for the technical fields. To a casual observer, political actions may appear to be inconsistent, capricious, immoral, and self-serving, and no doubt they are, on occasion, since intuitive judgments are made by fallible individuals, not automatons or "perfect" human beings.

The more one knows about how a political system operates, the more effective one can be in assuring more effective decision-making. This is as true in industry as it is in government. Because of the variability, each system must be observed and evaluated in a realistic manner if one is to be effective.

COMPROMISING AND WINNING

Two major characteristics of the political process are the art of compromise and the dedication to winning. While these characteristics are common psychological traits and not rare in other professions, they are dominant in the world of politics. The two characteristics are closely related, because compromises are necessary for a politician to win either reelection or support for his political convictions.

Rigidity, therefore, is not frequently encountered, nor does one find that a politician expects to "win them all." He is more likely to measure winning in terms of achieving a majority of victories rather than 100 per cent success. This apparent contradiction to the desire to win is allied to the politicians knowledge that he must compromise to broaden his base of support.

Compromise is not a failure in the eyes of most politicians. More than any other profession, politicians depend on obtaining a consensus which represents some position away from the two extremes.

While idealism is frequently evoked as a means of obtaining political support, it rarely (or never) produces an idealistic solution. Winning through compromise is a strategy that must be comprehended if one is to influence the quality of policy decisions.

CHALLENGING THE SYSTEM

"If you are not for me, you are against me" is not an uncommon attitude among politicians. Once a politician has publicly committed himself, any attitude other than full support is not appreciated. Objective reappraisals are not welcome, no matter how sincere the intentions nor how objectively they are conducted. Accordingly, the quality of the decision can best be influenced by analyses before the decision is made.

But frequently a decision needs to be challenged or at least reexamined. This reexamination will be most effective if conducted "within the system" wherein the decision-making politician has control of the results of findings. If conducted "outside the system," the results will be automatically rejected by the politican if they do not support his decision. Furthermore, those who support the decision-maker or who depend on his goodwill for their own well-being will also reject or at least not support any adverse findings.

Whether such attitudes are "right" or "wrong" is irrelevant. The decisions are matters of judgment, and no decision is either "completely right" or "completely wrong." The political system operates under the premise that too much priority on one set of values, or too little on another, will ultimately be decided by those who vote (in industry or in government).

Objective reappraisals from outside the system are construed as an attack unless the results are favorable. In addition, since an objective reappraisal is never desired, it is difficult to obtain funds for such studies.

Challenging the system is not an inappropriate activity for a scientist or engineer. The strategy to be used, however, must take into account the competitive character of such efforts, and must recognize that an antagonistic view will be taken by the political system.

FINANCIAL SUPPORT

In political systems, obtaining the goodwill of the decision-makers is critically important. "To the victors go the spoils" is considered to be a dominant characteristic of politics, but all persons in a position of power are accorded special treatment by those who are influenced by policy decisions. For family groups, business office relationships, industrial sales, and even the voter as he approaches the polls, the tendency to flatter, cajole, or "bribe" the decision-maker is a part of the game. In the business world, in and out of government, a part of the political strategy includes influencing the decision-maker or someone who has influence on the decision-maker. The methods used to influence individuals vary widely. The moral, ethical, and legal restrictions are also quite variable.

In the private sector, laws govern the extent to which one can "buy" the favor of an influential person. But because of its "private" nature, such activity in industry is of primary importance to the company involved, and of only general "moral" interest to the community at large.

For government employees, however, a public trust is involved; that is, all of society may be affected by the manner in which decisions are influenced. There are laws that define inappropriate behavior. There are also regulations or procedures that a given political system will enforce as means of protecting the public interest (or to avoid adverse criticism).

In general, the attitude of individual citizens is that government employees and those involved in elections to public office should not misuse their positions. On the other hand, the need for funds to conduct political campaigns had not been rejected in the early 1970s, and although federal laws are designed to limit such activity it is far from illegal or even unethical. As long as contributions to political parties are legal, it is inevitable that future decision-makers will be approached and funded by favor seekers—who subsequently hope to have at least a favorable reception from the successful politician.

The practice of contributing to political parties has been an obscure part of engineering activities. Although laws govern how such contributions can be made, they nonetheless are made, and the influence of an engineer on the political system is frequently related to how well he has supported a political party.

The extent (and even the desirability) to which an engineer supports political action is quite variable. The greater he concentrates on technology, the less is the importance of his involvement with the political system. On the other hand, the more he wishes to influence the lives of people, the more he may need to influence individuals in the political system.

Laws, regulations, codes of ethics, and individual moral principles ultimately govern the action of each scientist or engineer in his encounters with the political systems. To the extent that he is concerned about advancing civilization, to the extent that he is to be effectively involved with decisions that influence the "needs, desires and aspirations of man," he needs to understand the political system. Furthermore, he either needs to operate within that system, or he needs to change that system. Staying within the system permits immediate, short-range contributions. To change the political system, however, different strategies and longer time periods are necessarily involved.

Perhaps the most significant problem has to do with the use of hypocrisy and deceit as a strategy by technologists who become actively involved in the political system. In the final analysis, such activities are controlled by individual moral and ethical persuasions as a member of society. The desirability or necessity to invoke special "ethical standards" beyond those established by society is a political not a technological issue, based on the desire to change the political system, reduce the influence of the political system, or create a group with higher ethics than the norm. Suffice to say, the more a scientist or engineer is removed from the existing political system, the less influence he has on the quality of policy decisions.

THE ROLE OF GOVERNMENT

To govern, in the simplest sense, is to control or to exercise authority over the actions of others. Governments have been established by a variety of means and for a number of different purposes, and they inevitably limit individuals in some respects and enhance their opportunities in others. Governments can exist solely for the benefit of those who govern or they can exist for the benefit of those who are governed.

In the following section, the characteristics of government are examined to understand how government authority influences the allocation of resources. It is directed to the function of government rather than to the political processes, systems, or philosophy by which it operates.

Function of Government

In a society of free individuals, government has been described as a coercive institution that exists for the purpose of joint undertaking of tasks too large, too expensive, or too pervasive for single individuals to cope with. History provides a wide spectrum of types of governments, from those led by self-interested despots and tyrants to collective societies governed by the will of free and equal individuals.

Governments are coercive because of the differences that exist between individuals. There is a continuing need for compromise and for a sacrifice of individual choice and individual freedom. The measure of success of government and the ability of a political system to prevail is determined ultimately by the extent to which individuals accept the coercion. Thus a government that attempts too rapid a change, or imposes policies too extreme for individuals to accept, faces serious difficulties. If most individuals agree to the coercion, no problem develops. If they do not, greater force (and the resources to provide that force) is required and a myriad of individual decisions overrides the government action.

Anarchy—the absence of government and law—has few proponents in modern society. Disagreement does exist, however, on the extent and the character of government.

The need for government in a society of free individuals is simply to assist in the fulfillment of self-interest. Actions take the form of (1) protection of the individual's personal health, safety, possessions, and way of life, and (2) products and services that become available through joint undertaking of complex projects and tasks. Because of self-interest, individuals voluntarily subject themselves to (or accept) some form of coercion in order that the desired protection and/or benefits can be obtained.

The quality of life of individuals is influenced by the manner in which the government operates. However, for the purposes of this chapter, it is sufficient to recognize but not to debate the merits of alternative political and organizational characteristics of government.

In the United States, government consists of legislative, executive, and judicial branches. Simply, policy is set by the legislative branch; operations are directed by the executive branch; and the rights of the governed are protected by the judicial branch.

In the provision of engineering works, the policies that control technology come from legislative activity. The legislature is the group that must provide the type of policy decisions that can effectively utilize technology. The executive and judicial branches can clarify but not initiate the necessary action.

While the misuse of authority by the executive branch can lead to reducing effectiveness, the major burden for defining the use of technology utlimately rests with the lawmakers.

Three functions of government can be identified:

- Organization of certain social, economic, and political activity.
- Ranking priorities for society.
- Allocating costs and benefits to members of society.

Because society consists of both public and private sectors, three major collective choices must be made:

- The public/private mix. How much of the resources are to be managed by government?
- The make-up of the government activity. What will the government do with the resources that it controls?
- The present/future mix. What percentage of benefits is to be derived now, as opposed to that designated for some future time?

The ability of the government to influence the quality of life of individuals is in part dependent on the amount of resources that it controls, and the degree of autonomy with which it operates. Its effectiveness in achieving advancement can be measured by condisering its definition of what constitutes an improvement in quality of life.

Governments are a product of the political system which the individuals accept or to which they accede. The tasks of government increase in number, size, and complexity as population increases, as communities grow in size, and as wider and wider varieties of products and services are made available to individuals. As these joint tasks become more numerous and more complex, the power that is delegated to government must necessarily be increased. As a corollary, the degree of individual freedom is necessarily decreased. This trade-off between individual freedom and collective action has always existed. The benefits sought from joint endeavors have simply been a question of degree; that is, those who chose to live in urban areas have sought the material benefits that were possible through collective undertaking of tasks.

The ability of individuals or groups outside the government to produce changes which are deemed desirable is limited by the constraints imposed by government and by the available resources. The role of the private sector is discussed in a subsequent section.

Relative to obtaining effective use of technology, neither the form, the method, nor the political philosophy of government is a controlling issue. Technology can be and has been made responsive to many purposes. It has been used by many forms of government. Differences in government can and will result in differences in the benefits that are derived from technology.

An important distinction, however, is that differences in end results may well be due to differences in fundamental purposes of the government rather than to the inherent characteristics of technology. It is the criteria to which technology must respond that result in particular benefits—not the process. For example, the goal of industry and some governments is monetary profit or economic growth. In a free enterprise economy, the application of technology to economic goals has been studied in depth and is relatively well understood. The same can be said for other isolated criteria such as those for the Department of Defense and National Aeronautics and Space Administration.

Political philosophy and organization of government can and do influence the attitude toward, and the choice of, means to an end. *The type of requirements needed to control technological advance, however, are best known in the technological process and not from social-political variables.* The role of government is to provide the ways and means to solve the complex problems of a group.

In summary, the following principles of the function of government in a society of free individuals can be stated:

1. Government is a coercive institution that is required by any group of individuals who desire products and services that are too large or complex to be undertaken by individuals.
2. Government functions include the organization of social, economic, and political activity; the ranking of priorities for society; and the allocation of costs and benefits to society.
3. Governments must make three significant types of choices for society: how public and private activities are mixed; how public resources are used; and what benefits are to be sought in the present and in the future.
4. Governments are necessarily controlled by a political system that succeeds or fails depending in part on how coercion is applied and the extent to which the government satisfies the needs of those who are governed.
5. Since the political system controls government, the allocation of resources is also controlled by the political system.

Levels of Government

Each subordinate level of government is constrained from both above and below as to the extent of joint undertakings. Each level has some limits on its pursuit of objectives. But within its sphere of authority, the aforementioned three functions of government (organization, setting of priorities, and allocation of resources) exist as do the same three major collective choices (public/private mix, use of resources, and present/future mix).

At each level of government, problems of objectives and priorities must be resolved so that policy decisions are responsive to a specific set of priorities and are within the constraints on that governmental unit. Goals of government, particularly with reference to situations in which there is no apparent consistency or overall community-accepted set of objectives, are political, not technological, issues.

As stated previously, the question of how much control is retained by each level of government is a political and not a technical problem. Debates on this question cannot be resolved on technical grounds. Although inadequate understanding of technical problems can exist at one level of government, and although the retention of control by a higher level of government may produce better technical work, it does not follow that the public interest is best served by the loss of control by the subordinate unit. The quality of government includes not only superior technical work but also confidence and support from individual citizens.

As noted before, to obtain effective use of technology, requirements should be stated in operational terms for scientists and technologists. However, innovative solutions become increasingly more difficult to provide for the smallest segments of society. In small communities, the following things work against innovation: (1) the economics of production, (2) intercommunity relationships (for example, transportation), (3) political pressures for present rather than future gains, and (4) the political risks of introducing change when the status quo is more acceptable.

The special political problems that are present at a given level of government and that exist in a given community are critically important. The following principles apply to how the provision of technical work is influenced by the level of government:

1. Decisions to delegate authority and to establish rules, regulations and limits within which each level of government shall operate are the responsibility of the overall political system.
2. Each level of government is under citizen pressure to delegate authority to a lower level.
3. Each local level of government is constrained by limitations imposed by the political system.
4. Each level of government has the same basic functions and the same basic choices, but lower levels are controlled, in part, by higher levels of government.
5. Innovations are difficult to incorporate into solutions at local government levels.

Control Through Resource Allocation

As noted earlier, one of the three major functions of government is the allocation of costs and benefits to individuals in the society. This function, more than any other, controls the destiny of those who are governed. Whether the function is conducted in an orderly or disorderly manner is a social-political rather than a technical issue.

To fulfill the function of resource allocation, two major choices must be resolved by the political system. First, how much of the total resources will be used by government? Second, how will the importance of alternative undertakings be ranked; that is, which will have the highest priority?

With reference to the first of these, the entire range of resources of society is involved: (1) manpower, (2) money, (3) land, and (4) natural resources. Through the legislative process, the amount of resources that will be used for joint undertakings is resolved. The decision is made by the political system and is influenced by the form and autonomy of government.

The requirements for resources are ultimately controlled by the purposes of the society. The role of technology should be limited to advising government, or interested critics of government, on the consequences of the use of resources relative to society goals. In the United States individuals (or subordinate levels of government) strongly prefer to retain control over their own resources, which counteracts their desire to advance society through joint undertakings.

The second set of choices—setting of priorities—also is influenced by the degree of autonomy of the government. The more powerful it is, the less attention it will direct toward obtaining explicit approval of the rankings that are made. The setting of priorities faces the problem of (1) the short-term versus long-term benefits and (2) the potential benefit to individuals or groups of individuals. Banfield has classed a society that is willing to invest its resources for the benefit of future years or generations as a superior culture. Nonetheless, decisions to forego immediate benefits are not easily made, particularly in the face of opposition from vocal citizens.

The problem of who would benefit from government action did not pose too serious a question until the 1960s. Prior to that time, most policies were directed to achieving benefits for the "majority" or for an identified minority whose need was quite apparent. Generally, this also meant that some minority group (ethnic, property owners, neighborhoods, etc.) paid a higher price than did the majority, usually in the form of social rather than economic losses. The individuals who suffered these losses (individuals and groups as a result of benefits to a majority) became focal points for serious debates which led to a widening of the gap between the "haves" and the "have nots." These again, were political, not technical issues, and their resolution is a problem that has to be solved by political systems.

The question of how to deal with distributional effects to avoid wide discrepancies in life styles is one which requires a more precise understanding of social goals. Income distribution, public safety, health, education, recreation, racial and ethnic inequalities, and pollution of the environment became critical variables that needed to be evaluated, and priorities assigned for action. This is because the resources that are available for joint undertakings are always limited due to the aforementioned desire of individuals to control their own resources.

These are far from simple decisions, and they are made more difficult by the absence of a basis of comparisons. With only advocacy analyses available, the political system has only "muscle" operations for guidance; that is, those with the most political power are able to prevail. This results in an advocacy game, whereby policy development is largely intuitive.

The advocacy game was extended in the midtwentieth century in the United States as minority groups became aware of how the system worked. Furthermore, they were now able to be heard because society became more interested in life-enhancement. The result was a constant clash of political power and it became increasingly more difficult to produce a change in the status quo.

Difficulties are faced by the political system whenever (1) consensus on goals cannot be achieved, and (2) benefits from two or more widely different programs can not be compared on the same scale. Both of these problems come from the inability and/or unwillingness to classify the characteristics that describe well-being or life-enhancement. Given such a classification, there is an increased possibility that agreement can be reached as to which are the most critical.

Technologists are also at fault if the output of their works is viewed too narrowly or if the focus is solely on increased efficiency. While it is apparent that more efficient works advance certain aspects of society, it is *never* apparent (to the political system) how resources should be allocated.

As to control by resource allocation the following principles can be stated:

1. Unless there is a consensus on social goals, and unless it is possible to compare widely different programs on the same scale of benefits, the only alternative for the political system is to allocate resources to favor those with the most political power—the muscle game.
2. Where political power is the primary basis, the usual advocacy analyses fail to provide a comprehensive view of the consequences of a given allocation of resources.
3. Social welfare should be classified in a way that will facilitate obtaining a consensus on the needs of the community involved.

4. Technologists should analyze and provide results of how system outputs will be influenced by an allocation of resources, as a means of estimating the impact on the basic goals of society.

5. Technologists who prepare an advocacy analysis for use in resource allocation are participating in the political—not the technical—process.

Control Through Taxation and Regulations

One of the aforementioned functions of government is the organization of social, economic, and political activity. The intrusion of government into activities of individuals, organizations, or subordinate governments is resisted for the same reason as is the assessment of resources for joint undertakings. Taxes are the basis for accumulating monetary resources. Laws and regulations are used to control activities.

Taxation requires the solution to the problem of *who* should provide the means. This may or may not be related to who obtains the benefits. While analyses can be made of equity under various assumptions, the final decision of the political system will be on social-political not technical grounds.

Laws and regulations are an effective management tool of government. In the area of public safety, health, and the environment, laws and regulations are used to limit or constrain activities of individuals, groups, and subordinate levels of government.

The total impact of laws and regulations is obscure. The benefits that are sought provide the most visible impact. As noted earlier, however, the effectiveness of an action depends on how well it is supported by citizens.

The least apparent impact of laws and regulations is its function in resource allocation. For reasons of health, safety, environment, and conservation of resources, regulations produce either an increase in use of resources, or a reduction in the activities that are constrained. Thus various methods for achieving social, economic, and environmental objectives call for the allocation of more and more resources to satisfy overriding protective measures. For example, regulations to obtain safer vehicles result in allocating greater resources to highway transportation.

The indirect allocation of greater amounts of resources does not necessarily result from applying regulations. However, unless additional action is taken or unless the regulation produces a change in the desire or demand, the activity being regulated will usually use a greater share of society's resources. Classification of this indirect allocation of resources as "bad" or "good" is a social-political question. An important consideration in setting regulations is the impact that is made on resource availability as well as a quantitative understanding of the distributional effects. In the final analysis, laws and regulations constitute an allocation of resources and this consequence needs to be evaluated in light of other ways in which limited resources can be used.

For political systems, allocating resources by laws and regulations is frequently less difficult than collecting and allocating resources through taxation. However, actions are inneffective if not popularly supported and the result is (1) fewer benefits than anticipated and (2) a waste of resources.

Governments' choices on taxation and regulations are also related to the immediate and long-range goals of the society. To be consistent, taxes and regulations need to conform to the priorities that are established.

The following principles can be stated for the use of taxation and regulation:

1. Taxation is a management tool of government.
2. Distribution of benefits in accordance with taxation is a political not a technical issue.
3. Laws and regulations result in an allocation of private resources.
4. Laws and regulations should be evaluated in terms of the total impact on society, including the impact of the allocation of resources on the priority goals that have been set.

ROLE OF THE PRIVATE SECTOR

The private sector includes all nongovernment functions, ranging from activities of individuals and family groups to those of large corporations and privately owned utilities. The private sector, at least in an idealistic sense, has tended to be self-centered as compared to government; that is, concern for the public interests has been optional and secondary to individual or group objectives. Society may request but does not require that individuals or companies contribute to the general welfare beyond the paying of government imposed taxes, abiding by appropriate laws and regulations, and conforming to the moral and ethical codes of the community.

In principle, the governing or ruling groups within the private sector have the same functions and face the same choices as do the governments of society. The major difference lies in goals and objectives rather than in the principles by which activities are governed.

In the following section, the role of the private sector is discussed as it relates to the use of technology.

Function of the Private Sector

The private sector of a society consists of those activities conducted with resources for which the responsibility for management decisions rests with a private citizen(s). Within limits set by government, individuals and groups of individuals engage in the pursuit of a better life.

The activities of the private sector are in direct response to the needs and objectives of individuals. The collective interests being delegated to government, the private sector pursues courses of action that are of greatest importance to the individuals or groups that are involved. Priorities are placed in accordance with individual convictions and the size or extent of an activity is a function of the resources that are available.

The scope of alternatives that can be considered by an individual or groups is controlled by both the resources and the interests. The first requirement is survival, and then security. After these needs are met, the individual choices can range widely. Regardless of the political philosophy of the government, the problems of individual choice are the same and the attitude and conviction of the individual(s) provides the basis for decisions. One of the most difficult choices is the extent of the efforts that will be directed to the development of personal resources. This is the same type of problem encountered in government wherein the level of resources and the present/future mix choices must be made.

The larger the community to which an individual chooses (or is forced) to belong, the greater the constraints that are imposed. The more removed from society that an individual chooses (or is forced) to be, the greater is the freedom from limitations imposed to protect joint interests, but it also reduces the number and types of alternatives for the pursuit of a better life.

An individual faces most of the same type of problems that his government must face. First, there is the problem of how much resources he will (or can) try to obtain to further his objectives. Second, given a certain level of resources, he allocates them in some manner. But he must also face choices about the present versus the future and of benefits from joint social undertakings that are voluntary but of little direct benefit to himself.

Perhaps the most significant reason for considering the role of individuals is the clarity that it brings to considerations of other government and private sector activities. Problems that an individual faces in charting his own path or that of his family are quite similar in principle to those faced by larger units. An extension of the problems of individuals to the problems of nations will provide considerable insight on the principles that are involved.

This is *not* to say that the problem-solving by an individuals can be simply extended to solve problems of large communities. There are tremendous differences in scale, as well as theory of behavior, in decision-making by government.

Businesses and corporations grow out of a decision by a group of individuals to conduct joint undertakings too large for a single individual. These joint undertakings require the same type of sacrifice of autonomy as does the similar delegation of authority to government. A major difference is that individuals are free (idealistically) to disengage themselves from one company in the private sector and join another. Another significant difference is that the public interest is not necessarily a consideration.

Most joint undertakings in the private sector are for one objective—to increase the resources of the individuals involved. Exceptions include the quasi-government operation of foundations, religious groups, and cultural programs that are funded by those who choose to donate resources in a particular manner. These nonresource development activities are provided in the public interest, and, because of their objectives, they are quite comparable to government.

The objective of most joint undertakings of the private sector is for private monetary gain. However, the activities provide other vital service to society. Many different products and services are made available which could not exist except through joint undertaking either public or private. Most large businesses and corporations are funded by individuals who in turn receive the profits that are derived. If the product or service is needed but profit cannot be derived through individual investments, funds (subsidies) are provided by government.

In summary, one of the functions of the private sector is to provide the means by which products and services are made available. Not supported by government grants, companies must produce a profit or they cannot continue producing. Contributions to society other than the production of goods and services is not possible unless the increase in costs to provide additional benefits can be recovered through higher prices.

The following principles may be stated relative to the private sector:

1. The private sector consists of activities managed by private citizens with private resources.
2. The objectives of private sector activities are set by the private citizen(s) involved with the management decisions.
3. One of the major functions of the private sector is the provision of products and services.
4. One of the major requirements of the private sector is that the activity must provide a profit or at least operate within its level of resources.

5. There is no requirement that the private sector contribute to the public good beyond the actions set by public governments.

Operation of the Private Sector

The primary social responsibility of the private sector is to produce goods and services (physical resource development). As noted in the preceding section, it does not have the responsibility of providing for the general welfare or social well-being of individuals. An individual business or corporation may choose to do so, but this alternative may well exist only under noncompetitive conditions. Under a free enterprise, market-place economy, the private sector will produce only if a profit is derived by those who finance the operation. Under political systems in which production is controlled by government, profit is not required and social well-being can be incorporated into any manufacturing or service function.

The conduct of business in a free-enterprise economy is constantly faced with the problem of risks. On the production side, these risks are in the form of over-or-under-production, loss of resources that are invested, and similar problems related to economic survival or desirability. Success in the market place depends upon how these risks are resolved.

A different type of risk is related to the social-political problems of the private sector. These problems have to do with the safety and well-being of individuals in the community. Pollution, neighborhood deterioration, and similar adverse effects are never completely eliminated, and if there are no government regulations, a company must take the risk of damaging the community or loss of competitive position in the market place.

Equally troublesome are the risk questions of product (or service) safety. Practically any product or service can be misused, or can be considered "unsafe." Furthermore, the manufacturing and service functions are designed or performed by imperfect human beings which means that some errors are inevitable. Perfection is an Absolute (in philosophic terms) that cannot be achieved. The closer that perfection is approached, the greater the cost.

Closely allied to how "perfect" goods or services should be, is how much risk an individual is willing to take. As products and services become more complex, the ability of individuals to determine risks becomes all but impossible. As a result, governments set regulations that prevent the sale or use of unduly risky items or services, or require that the product or service clearly identify relevant characteristics.

If a component of the private sector undertakes to protect the public interest, it becomes a part of the political system or government of the community. This is inevitable to some degree. How much the private sector should be concerned with "governing" the community is obviously a political issue in itself.

Under the free enterprise system, imposition of regulations by government is one way that the public interest is protected. To some extent, public opinion can control the actions of the private sector, but with complex systems, this approach is less effective than government action.

Within the framework of the private sector, technologists must provide facilities and participate in the production of goods and services. As a technologist, he can predict the performance of various subsystems under various types of inputs, including the total consequences of the technical works.

The following principles can be stated relative to operations in the private sector:

1. The operations of the private sector are controlled by top managements' resolution of the following types of risks, relative to achieving their primary objective of being a profit-making activity:

- Over versus under production
- Loss of investment
- Unsafe and unwholesome community environment
- Unsafe products or services

2. Technologists should predict changes in performance of subsystems.
3. Technologists can influence management decisions through (a) developing a knowledge of the social-economic-political problems facing the organization, (b) use of the political system that controls the organization, and (c) public political action.

Government Constraints

Limits on the activities of the private sector can be imposed by governments for four types of protection of the public interest;

- Health and safety
- Consumer protection
- Conservation of resources
- Social stability

The first two are concerned with day-to-day well-being of individuals, while the latter two have both long-range and short-range implications for society.

The constraints to conserve resources and for social stability have the major effects on the quantity of output, whereas the first two, concerned with individual welfare, influence the quality of output. Imposition of these constraints is a responsibility of government (political system) and is resolved by social-political judgments.

Constraints are imposed to offset an adverse impact. What is not apparent from the imposition of a constraint is the changes that are produced in various subsystems. These subsystem changes can produce unexpected impacts throughout society.

The problems generated by ill-conceived constraints result from attempts to control a single impact on individual well-being. But ultimate impacts are a product of many activities and a limitation that is imposed on one impact will set off a chain reaction that can produce unexpected and undesired impacts.

It should also be noted that constraints on several single impacts usually result in a conflict of "ultimate values." This leads to confusion and inconsistencies (e.g., more than one ultimate exists). Examples of government constraints that are competitive and inconsistent include clean environment and safety which are mutually exclusive benefits in many instances. In most areas of risk, the threat of personal injury or death can override the availability of products (or services) that some individuals wish (e.g., fireworks and smoking).

It is not that government constraints are not needed in society; rather, the analyses that are used and the manner in which the constraints are imposed are important considerations. The objective of constraints is to limit the seriousness of ultimate impacts. By analyzing the subsystems that produce this impact, the alternative actions can be identified. This may lead to control of subsystem outputs in a manner that permits the resolution of conflicts between mutually exclusive benefits.

In the final analysis, if the "ultimate value" is the quality of life of individuals, constraints that are imposed by government should not inadvertently reduce that quality. Furthermore, the trade-off in benefits should be understood to the extent possible.

The following principles are relevant to the question of government constraints on the private sector:

1. Constraints on the private sector are imposed in the public interest by the government to:

- Protect health and safety
- Protect consumers
- Conserve natural resources
- Achieve social stability

2. Constraints that are imposed to reduce one adverse impact on individual well-being will usually result in a worsening of another effect.

3. The setting of a series of disconnected constraints leads to conflicts when two or more of the desired impacts become mutually exclusive.

4. Technological analyses of changes in subsystem performance can provide insight on the consequences of imposing constraints.

SELECTED READINGS

8.1 Albertson, M. L., L. S. Tucker, and D. C. Taylor (Eds.), *Treatise on Urban Water Systems*, Colorado State University, July 1971.

8.2 Baram, M. S., "Technology Assessment and Social Control," *Science*, Vol. 180, No. 4085, 1973.

8.3 Bauer, R. A., *Social Indicators*, Cambridge: MIT Press, 1966.

8.4 *The Behavior and Social Sciences, Outlooks and Needs*, Englewood Cliffs, N.J.: Prentice-Hall, 1969.

8.5 Bell, D., "The Idea of A Social Report," *The Public Interest*, Spring 1969.

8.6 Curtis, M. (Ed.), *The Nature of Politics*, Avon, Hearst Corporation.

8.7 Foss, P. and D. Hill, *Politics and Policies: The Continuing Issues*, Wadsworth, 1972.

8.8 Lerner, D. and H. D. Lasswell (Eds.), *The Policy Sciences*, Stanford University Press, 1951.

8.9 Olson, M., Jr., "The Plan and Purpose of A Social Report," *The Public Interest*, Spring 1969.

8.10 Snow, C. P., "Science and Government," *Science and Technology*, January 1969.

8.11 Sorensen, T. C., *Decision-Making in the White House*, New York: Columbia University Press, June 1963.

8.12 Wade, L. L. and R. L. Curry, Jr., *A Logic of Public Policy: Aspects of Political Economy*, Wadsworth, 1970.

CHAPTER 9

THE ROAD AHEAD

Policy development is the process for analyzing how resources are to be allocated. The result of the policy development process is a set of alternatives on which a decision on use of resources can be made. The success of decision-making is *absolutely dependent* on (1) the adequacy of the policy development process and (2) the decision.

Policy development can consist of very sophisticated analyses or of simple, instinctive reasoning. It can be a process for weighing public interest, or it can be a technique that simply weighs self-interest. It is a process used by every level of government, every private company, and institution, and by every individual.

Policy development is not decision-making. Neither is it planning, administration, budgeting, design, construction, maintenance, or operations. Policy development and policy decisions precede these processes.

The distinction between the two processes can be seen by a study of Tables 9.1 and 9.2. As Ogden has said: "Good engineering plans and good

Table 9.1 Political Process for Allocating Resources

Step 1	Decision on criteria (value system to be used).
Step 2	Perception of need (needs and desires of individuals).
Step 3	Evaluation of resources (uses, generation, and allocation possibilities).
Step 4	Evaluation of alternatives (competition and compromises).
Step 5	Decision on goals (setting priorty).
Step 6	Decision on means (feasibility).

economic analysis can determine whether or not development is possible. Politics decides whether it will be done."[1]

Policy development is as old as man, dating back to the first time that there were alternative ways for a man to use his resources. Over the centuries, decision-making on national resource allocation has become more complex, and the analyses that precede a decision have become increasingly more sophisticated. Nonetheless, decisions, by definition, require more than mathematical manipulation—that is, the answer to a mathematical problem is called a solution, not a decision.

Questions on how resources should be allocated will not grow less complex in the future, and will not disappear as long as individuals and their governments have any variability on what constitutes a better life. Accordingly, the process of policy development will be continuously improved. Political philosophy can change the *basis* for decision-making, but it cannot eliminate the process herein called policy-development.

Table 9.2 Technical Process Related to Allocation of Resources

Step 1	Develop quantitative descriptions of existing societal systems (quantity and quality of service).
Step 2	Develop quantitative descriptions of the resources used and the indirect effects of existing societal systems. (All resources and all nonfunctional effects).
Step 3	Predict changes in quantity and quality of service and indirect effects of societal systems that will be produced by changes in resources used, constraints, or internal operations.
Step 4	Predict impact on societal criteria of changes in quantity and quality of service and indirect effects of societal systems.
Step 5	Evaluate alternative means to achieve societal goals.

[1] Selected Reading 8.1.

The specific changes that knowledge will impart to fulfill the future needs and desires of human beings are difficult to predict. Certainly, however, science and technology will have a limiting or constraining influence, regardless of what form of culture is sought. It does not matter whether a society chooses to go forward, backward, or sideways, or by how these directions are measured, success will be achieved in proportion to how effectively science and technology are used.

The preceding chapters have attempted to describe, define, and develop the process of policy development. In this final chapter, the earlier concepts are re-examined and somewhat amplified to facilitate development efforts and broaden applications.

SUMMARY OBSERVATIONS ON THE POLICY DEVELOPMENT PROCESS

Precise Policy Goals

The need for the process of policy development grows out of two conditions that cannot be resolved by reasoning or logical processes:

A conflict in values. For example, which of such diverse values as those related to health, education, esthetics, and such, should be given highest priority when questions on resource allocation must be answered?

The need for compromise. That is, how much should programs be reduced for economic growth, conservation of natural resources, reduction of pollution, increase in public safety, and such, in order that some other program can be maintained or increased?

The results of policy development lead to two separate, but related decisions:

That a specific goal or benefit will be sought with a specified level of resources; that is, a program should be funded.

That a specific method will be used to achieve the benefit; that is, *how* the goals are to be met is specified.

The first decision needs to be specific as to the *quantity* of the change (or benefit) desired and the amount of resources that will be allocated. The second decision needs to be specific as to the *quality* requirements for the method that will be employed.

The quantity-quality-resources complex defines and limits the scope of the goal. In fact, if two of the three are specified, then the feasible range of the third is quite restricted. For the most effective decision, a balance must be reached between a too rigid and a too nebulous statement. If overly restrictive, then good alternatives may be inadvertently eliminated. If too flexible, then an infinite number of solutions are available, and limits will most probably be set by the analyst.

Insight on how to achieve a balance between quantity, quality, and resources will require technologists who are capable of estimating the performance of both technological and social systems.

Guidelines for Policy Development Studies

Based on the definitions and discussions in earlier chapters, the following guidelines for the conduct of policy development can be stated:

1. Policy development, by definition, must precede the decision-making. In fact, the results will be of little value (except for strategy purposes) if a decision has already been made.
2. A nonadvocacy analysis should be conducted; that is, the report should produce objective information for the decision-maker, rather than arguing for the decision that the analyst would make if he were the decision-maker.
3. The decision-maker should establish ground rules for the conduct of policy development to avoid obtaining slanted, biased, or incomplete information.
4. The adequacy and effectiveness of the policy development process is largely dependent on how well the following three sociopolitical issues are resolved:

- Classification, measurement, and evaluation of the quality of life of individuals and groups.
- Body of theory for predicting the impact on these variables (for measuring life qualities.)
- Political strategy for compromising power struggles between adherents of mutually exclusive values.

Advocacy reports can be useful as a part of the decision process or as a part of the political strategy for implementing a decision. If nothing but advocacy reports are available prior to a decision, information on some solutions may not be available. This is particularly true for innovative solutions.

Linking the Technical and Political Processes

The process for deciding how to use one's resources (individual or collective) involves the six steps listed in Table 9.1. That is not to say that such a rational, consistent or continuing process is followed. But consciously or unconsciously, these issues are present and are resolved.

The allocation process requires four subjective preferences; that is, only steps three and four can be accomplished on an objective basis. As noted in earlier chapters, it is steps one, two, five, and six that require action from the leadership of the political process. Though advice may be sought (or received), the political leadership cannot escape the responsibility for and the consequences of making political judgments.

Relating the political to the technical process is a major objective of the policy development process. Table 9.1 contains two steps that call for technical analysis, while Table 9.2 lists the activities that require technical competency.

The policy development process, while consisting of the contact point between the political and technical processes, is under the political process—that is, the latter controls the way that the analysis are conducted. Otherwise, the results could well be worthless. On the other hand, the *results* must not be controlled, for under these conditions, a decision will in effect have been made, and the study is probably a part of the political strategy for implementation.

Because policy development includes technical analyses, the process can start either at the decision-making level or at some lower level. To be effective, however, the studies need to be useful to the decision-maker. Therefore, the criteria and priorities by which the results are to be weighed need to be those of the decision-maker. This means that functional criteria need to be converted to societal criteria in order that comparisons of benefits between quite different activities can be made.

Table 9.3 suggests a type of procedure for setting societal goals, while Table 9.4 is for setting functional or activity goals. The necessary relation to societal goals are indicated. Policy development related to setting functional goals consists of evaluation of alternative means for achieving societal goals.

PERFORMANCE OF SOCIAL SYSTEMS

The performance of social systems will have measures of both physical and psychological variables. These measures are less difficult to identify than to measure or to rank.

The performance is largely controlled by the political philosophy under

Table 9.3 Typical Procedure for Setting Societal Goals

Step 1	Establish existing data base.
Step 2	Establish perception of needs.
Step 3	Establish existing societal system criteria.
Step 4	Identify existing policy (budget).
Step 5	Estimate future effect of existing policy (on social criteria).
Step 6	Obtain political commitment to priorities for improvement in societal criteria.
Step 7	Obtain political commitment to immediate and long range goals:
	a. Quantity (elements and modifiers)
	b. Resources (all types)
	c. Constraints (impacts on criteria of step 3)
	d. Time requirements
Step 8	Establish alternative means to achieve goals.
Step 9	Obtain political commitment to specific means.
	a. Quantity (of means)
	b. Resources
	c. Quality criteria
	d. Constraints
	e. Time requirements

which the social system operates. Figure 9.1 illustrates the approach of a political system which is based upon achieving a society of "Free and Creative Individuals." For each and every level of government, it will become increasingly more essential to resolve the political issue of how success and failure of the social system is to be stated and measured.

Certainly during the 1970s and 1980s it is important to understand and to develop an operational understanding of what has been described in Chapter 7 as the community culture.

As noted in earlier chapters, there have been efforts to develop social accounting through the use of *social indicators*, or measures that would parallel economic indicators. These barometers were to provide a basis for measuring individual well-being in nonmaterial terms. In the following section, two approaches to the classification and measurement of performance of social systems are discussed for the state of the art in the mid-1970s.

Social Indicators

As noted earlier, social indicators are useful for social accounting and as a means of discussing and measuring the many characteristics that influence

Table 9.4 Typical Procedure for Setting Functional Goals

Step 1	Identify existing data base. a. Quantity b. Quality c. Resources
Step 2	Identify functional criteria.
Step 3	Identify impacts on societal criteria.
Step 4	Identify perception of need.
Step 5	Identify existing policy (budget).
Step 6	Identify alternative functional goals and their impact.
Step 7	Establish functional goals (short and long range). a. Quantity b. Resources c. Quality (criteria) d. Constraints (societal criteria) e. Time requirements

an individual's view of how well his needs and desires are being fulfilled. A comprehensive assembly and organization of data for social indicators was released by the federal government in 1974. The introduction to these data stated:[2]

> Ideally, an indicator would show, in a timely fashion, the status of the population in relation to a particular concern. It could be disaggregated to show which groups of the population were affected, and it could be linked statistically with other indicators to relate change in one condition to change in another. Thus, an indicator would reveal not only the status of the population in relation to a perceived social objective, but it would also furnish some ideas of what forces were influencing that status. At the present time, not enough is known about the cause and effect of social conditions to develop such ideal indicators. Rather, the indicators presented in this publication represent simply a first step toward the development of a more extensive social indicator system.
>
> The choice of indicators is based upon two main criteria: That the indicators measure individual and family (rather than institutional or governmental) well-being and that they measure end products of, rather than inputs into, social systems. In education, for example, the indicators were selected to measure individual achievement and attainment rather than inputs, such as school budgets, classroom construction, and the number of teachers.

In effect, social indicators are a classification and measurement technique for the performance of a social system. The eight categories in the

[2] Selected Reading 9.8, p.xiii.

referenced report are listed in Table 9.5. The first seven categories describe *qualities* and the eighth is concerned with *quantities* of a social system. Figures 9.2 to 9.7 are typical charts from *Social Indicators, 1973*.[3]

The use of social indicators in policy development requires a good understanding of both the significance and limitations relative to their (1) purpose as *indices* rather than objectives, (2) *political significance*, and (3) *variability*. Like temperature and weights, the level of an indicator is not

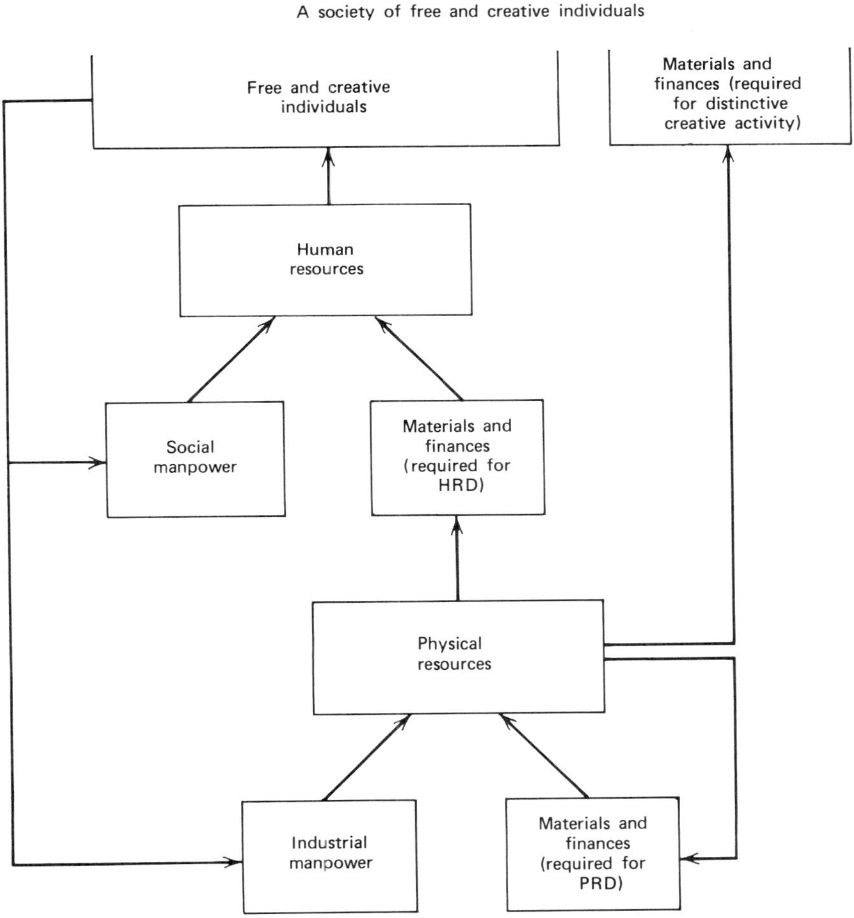

Figure 9.1. Total resources development.

[3] Selected Reading 9.8.

Table 9.5 Classification of Social Indicators[a]

1. Health
   a. Long life
   b. Disability
      (1) Long term—institutional and noninstitutional
      (2) Short term
   c. Access to medical care
2. Public safety (safety of life and property from crime)
   a. Violent crime
   b. Property crime
   c. Freedom from fear of crime
3. Education
   a. Basic skills—attainment and achievement
   b. Higher and continuing education
4. Employment
   a. Employment opportunities
   b. Quality of employment life
      (1) Job satisfaction
      (2) Working conditions
5. Income
   a. Level of income
   b. Distribution of income
   c. The low-income population
6. Housing (housing quality)
   a. The housing unit
   b. Living space
   c. The neighborhood
7. Leisure and recreation (leisure time)
   a. Outdoor recreation
   b. Television viewing
8. Population
   a. Growth
   b. Distribution (geographic)

[a] From Selected Reading 9.8.

sufficient in itself to establish relative quality. For specific conditions, temperature and weights do have ideal levels. But obviously, for example, the ideal temperature for refrigerating meats is quite different from that for cooking meats.

While some social indicators, such as health (Figure 9.2) and public safety (Figure 9.3) may well have an ideal condition, others such as housing

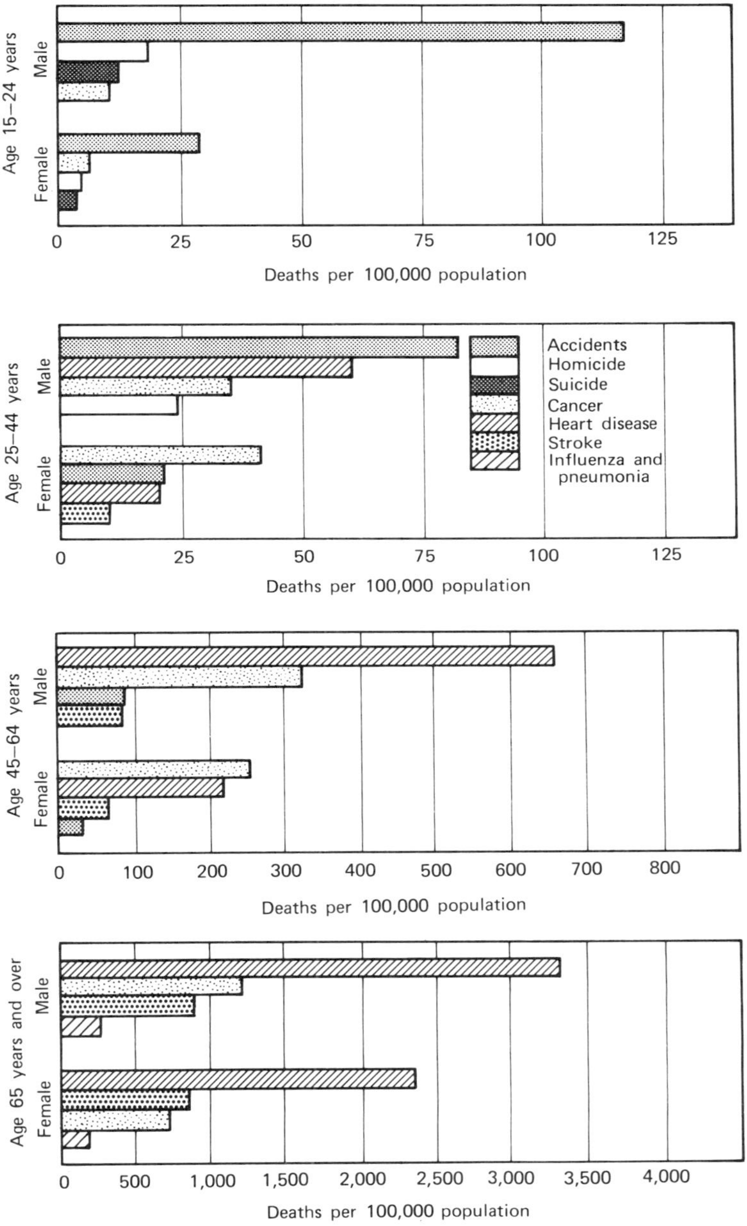

Figure 9.2. Death rates—four leading causes.

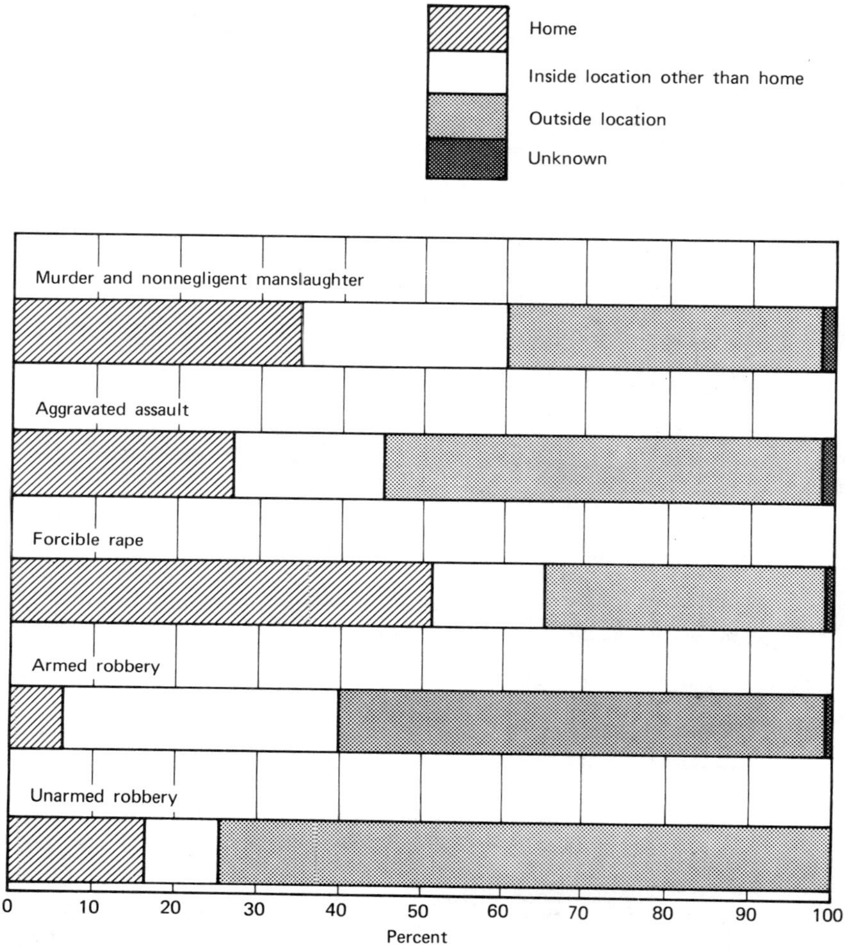

Figure 9.3. Place of occurrence of violent crime.

(Figure 9.4) and recreation (Figure 9.5) will vary with individual likes and dislikes. Furthermore, the output performance of a social system is a function of the input, and one of the inputs is economic resources. Since the latter is limited, as is technology, the ideal condition in terms of material well-being for a social system is a balance between what is desired and what can be achieved for a given economic and technologic input. Therefore, the social indicators need to be viewed as being basically amoral, because any

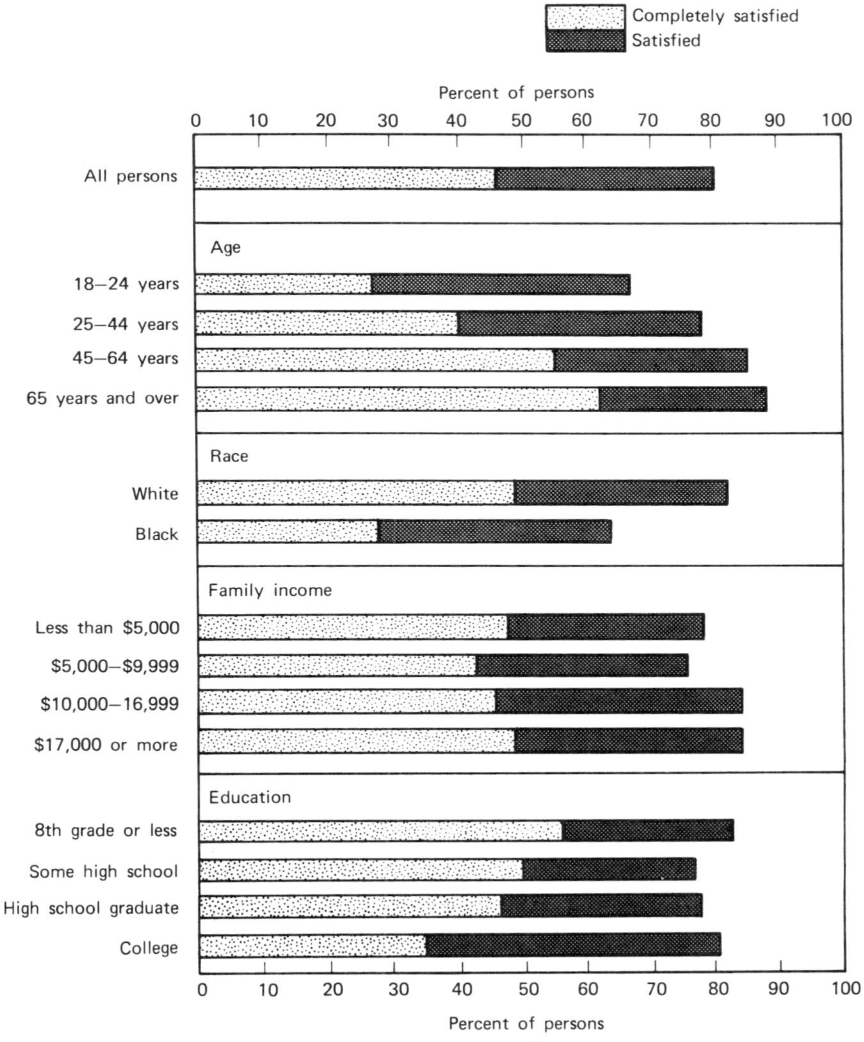

(a)

Figure 9.4. Housing. (*a*) Satisfaction with neighborhood. (*b*) Households living in substandard units.

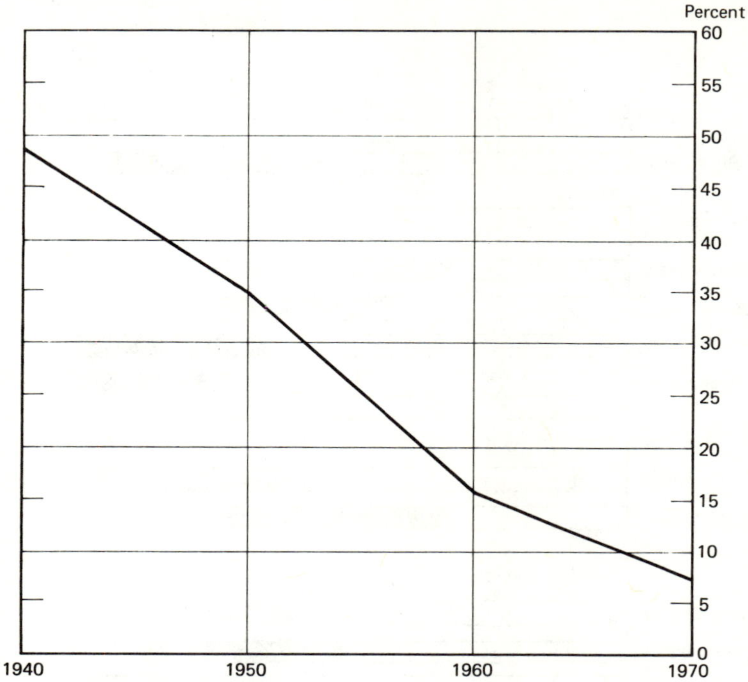

(b)

Fig. 9.4. (*Continued*)

judgment as to the necessity for improving an indicator requires a judgment that another indicator can be neglected, or at least can receive minimal adjustment.

Changes in levels of social indicators may well become short-term or long-term goals. If so, the major problem is how the changes should be brought about, and it is to this question that policy development is addressed. Many activities contribute to improving an indicator, but frequently result in the degrading of another indicator. In fact, more times than not, an improvement in one indicator produces an undesirable change in another. Such a result is inevitable, since there are (1) limited resources and (2) conflicting values.

The *political significance* of social indicators is related, in part, to the resource allocations that are necessary, and the setting of priorities that cannot be avoided. Decisions are constantly required for setting laws and

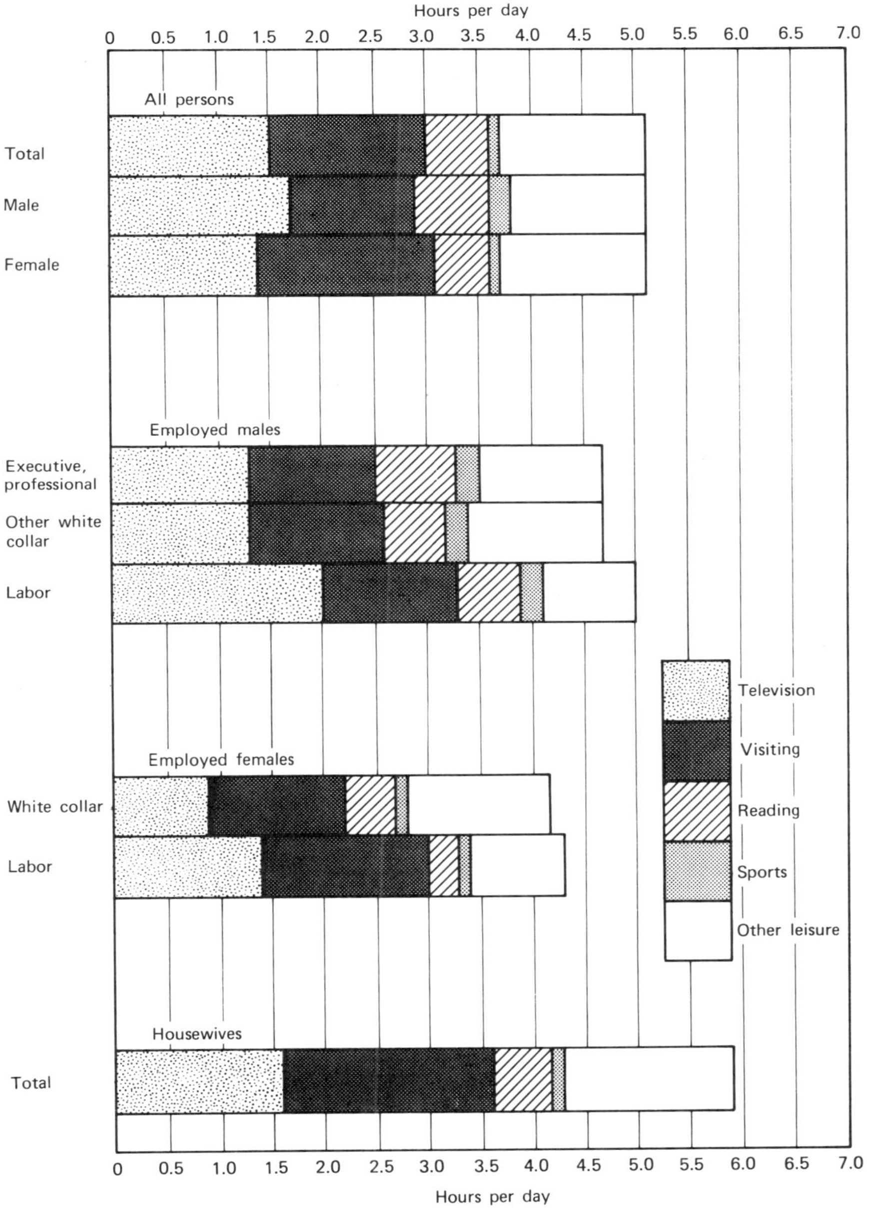

(a)

Figure 9.5. Leisure and Recreation. (*a*) Daily use of leisure time—1966. (*b*) Participation in outdoor recreation.

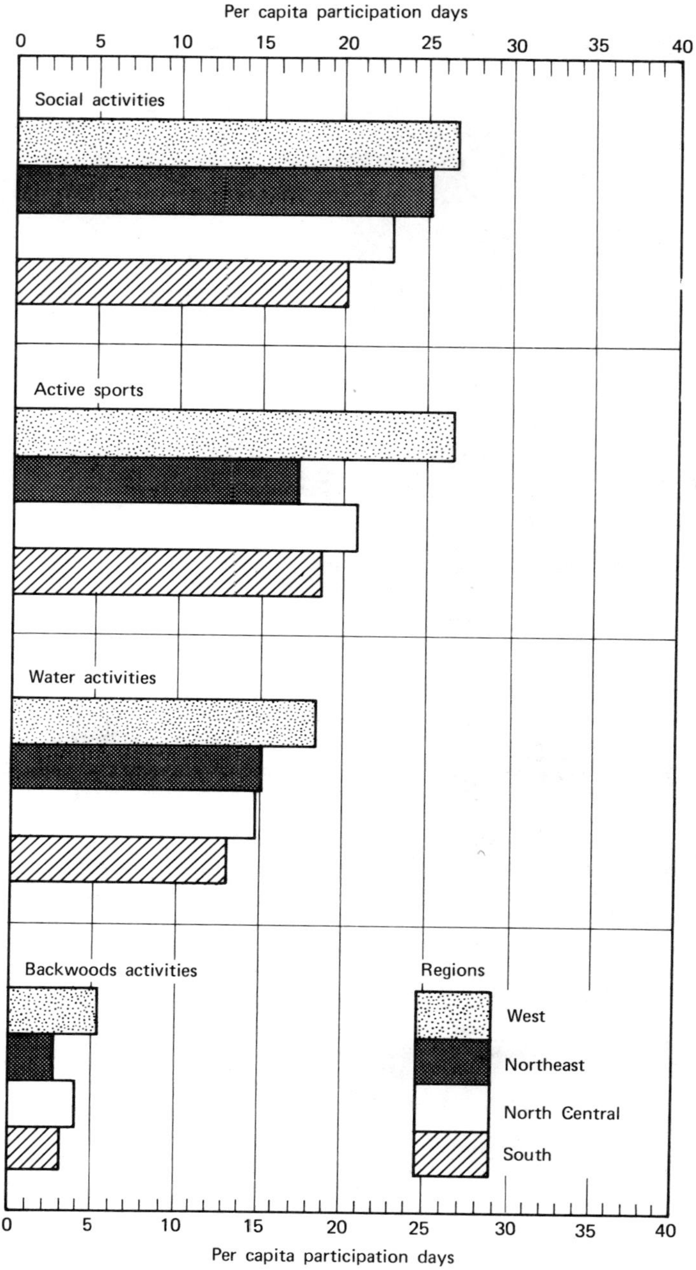

(b)

Fig. 9.5. (*Continued*)

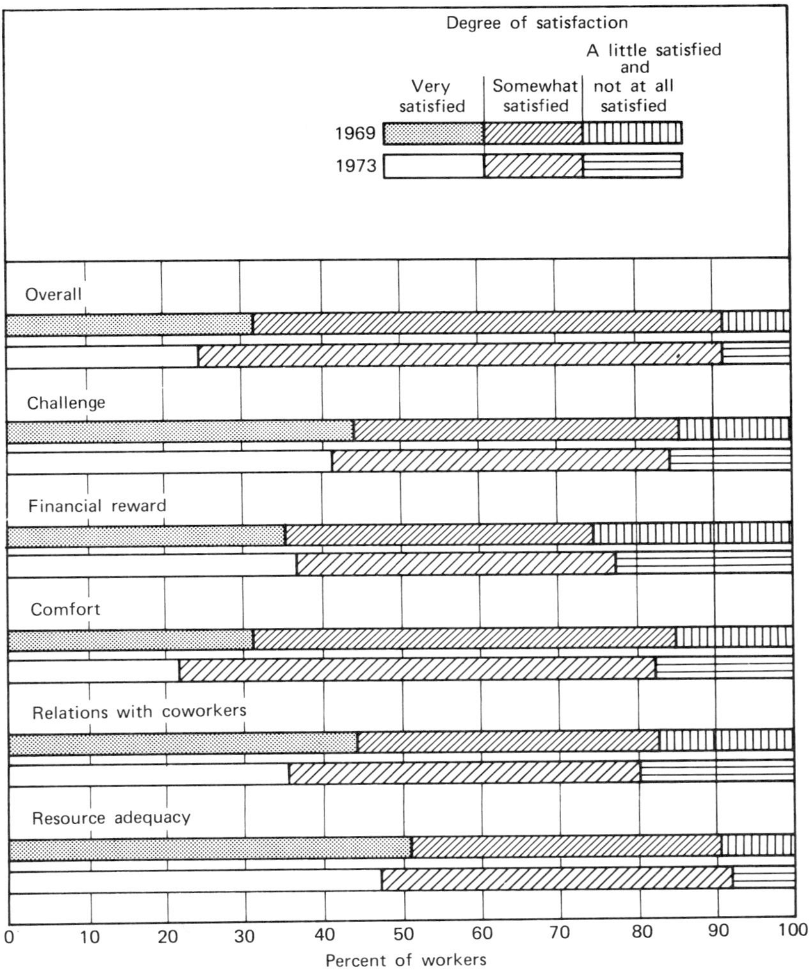

Figure 9.6. Job satisfaction.

regulations and for budgeting public funds to various activities. Each of these actions is concerned with improving the performance of the social system, no matter how that performance is defined or measured. If social indicators are to be used by the political decision-maker, they must provide information that is relevant to his perception of social needs and desires. To provide data on simply improvements of functional efficiency of a subsystem may be quite useless. For example, the fatality rate of highway

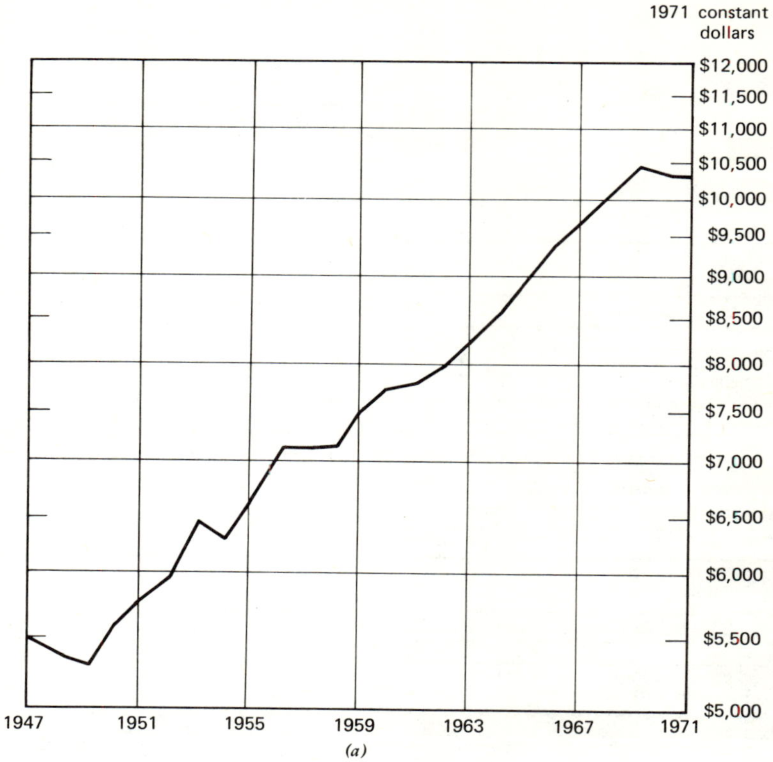

Figure 9.7. Income. (*a*) Median family income. (*b*) Consumer unit expenditures. (*c*) Profile of the low-income population.

deaths based upon vehicle-miles of travel may not be very helpful. The number of school buildings, the size and location of hospitals, and the acreage of wilderness areas may also be of little real assistance to the political process.

Social indicators are of assistance in policy development only if they are in forms that mean something to the decision-maker. This requirement is particularly critical for smaller communities where the perception of need can be more directly related to functions or activities. Accordingly, the political system of the community, large or small, ultimately controls whether social indicators or some other basis is used to measure the performance of its social system.

The *variability* of output is an important characteristic, no matter how

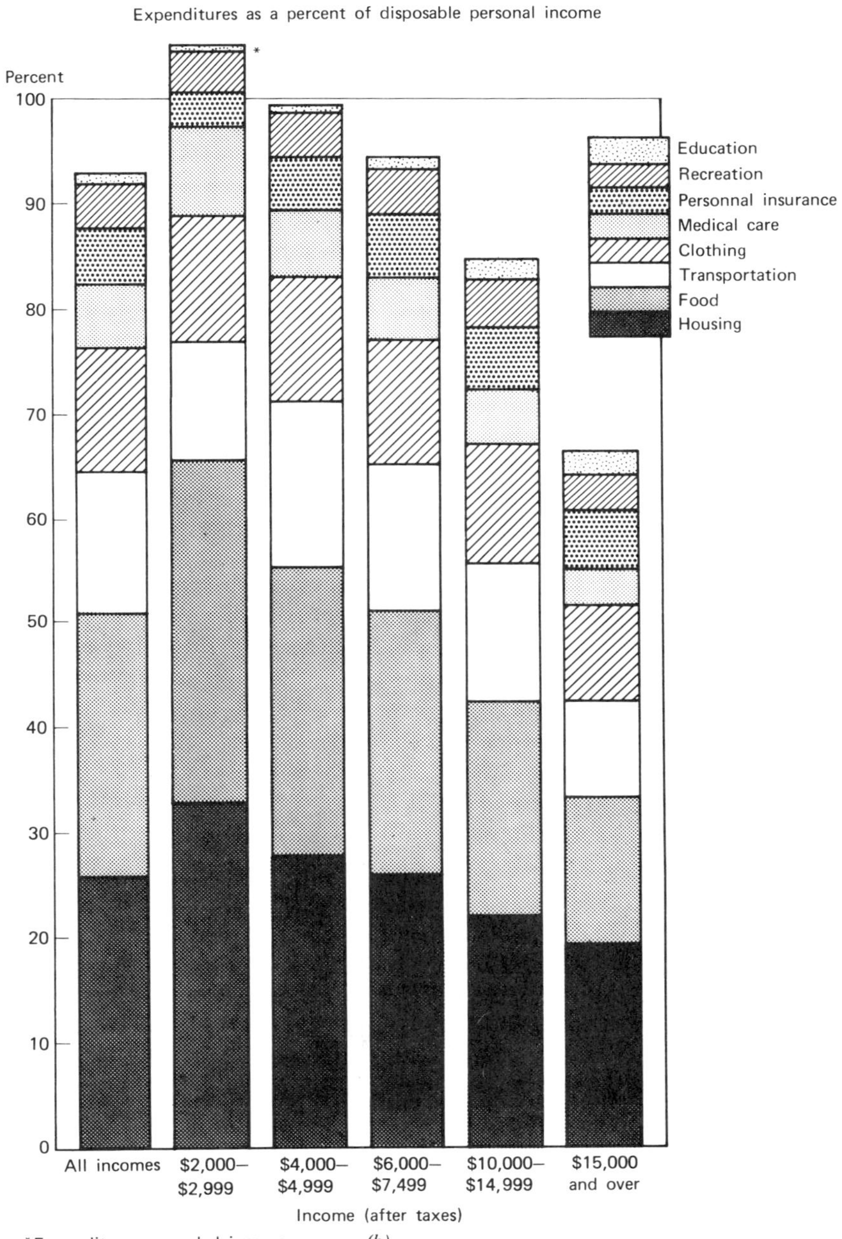

(b)

Fig. 9.7. (*Continued*)

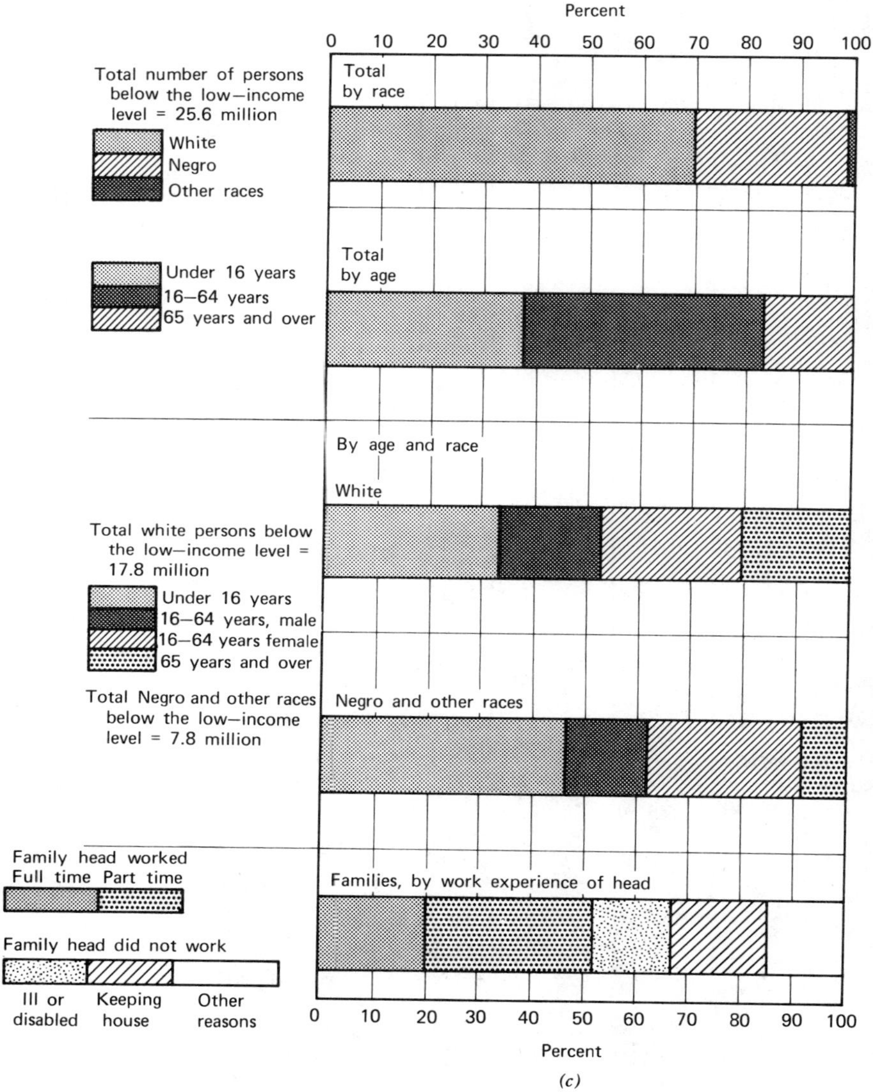

Fig. 9.7. (*Continued*)

the output is classified. The earlier referenced report disaggregated the data to show the variability. Relative to disaggregation, the report stated:[4]

> In almost every case, the national totals are disaggregated to show at least the age, sex, and racial characteristics of the population. The degree of detail varies from indicator to indicator and, wherever possible, has been determined by identifying those groups in the population which are most vulnerable to, or most affected by, a particular condition or which present a unique situation.
>
> The main reason for this disaggregation is to identify and compare significant groups within the population and to show their changing conditions relative to each other and to the national average.

Three types of variability can be identified: normal, temporal, and geographic. The normal variability refers to the usual differences (Figure 9.2) between individuals that exist within a population. Part of this variability is due to actual differences and part to measurement error. Temporal variability refers to the changes from one time to another in the conditions (Figure 9.4*b*) and in individual perception of conditions (Figure 9.6). The latter is not valid as an objective measure of conditions, but can be used to identify the perception of need. The geographic variability refers to the differences that exist due to climate and other characteristics that are influenced by the geographic location of the individual or the activity (Figure 9.5*b*).

Societal goals need to account for variability. Thus for improvements in health, will the objective be focussed on increasing life expectancy? If so, which groups in Figure 9.2 should be influenced? If violent crime is to be reduced, which of the variables in Figure 9.3 should receive priority? If level of income is a problem, should efforts be directed to changing the median income of Figure 9.7*a*, in the spending habits of Figure 9.7*b*, or should the low-income group (Figure 9.7*c*) be assisted? The question of *how* resources should be allocated will be dependent upon which segment of the problem is to receive highest priority. If either a shot-gun approach or an intuitive response is made to a perceived need, the effect on the performance of the social system is likely to be quite different than if an objective appraisal of variability precedes the decision.

Perception of Need

One alternative approach to social indicators, which can be readily related to the perception of need, identifies the most serious problems that exist.

[4] Selected Reading 9.8, p. xiii.

While simple to state, the task of determining which of the existing needs are most critical poses difficult political problems. Actions that are taken that are based upon critical needs are likely to have popular support. It will be difficult, however, to explain to those with other needs why their interest is not being funded.

A hazard with this practical approach is not only in the difference in priorities that exist in society. It also leads to the treatment of symptoms rather than causes.

At levels other than the federal government—state, county, municipal, rural, or the private sector—the problem approach is particularly useful. Here the critical needs are quite evident, and the use of social indicators makes less sense than does the concern for specific flaws.

The following guidelines for dealing with performance of social systems can be stated:

1. The quality of life needs to be classified, measured, and evaluated to obtain the most effective policy decisions; without such a classification, it is difficult to offset or minimize the deleterious effects of political power as the sole basis upon which decisions are made.
2. A single index of the quality of life will not usually contribute a useful basis for political judgement. Therefore, a spectrum of the impacts, such as the effects on social indicators, will be much more descriptive of the benefits and adverse effects of alternative goals.
3. Policy development should proceed under a classification system for the quality of life or a priority of problems that is approved by the decision-makers. This system can be (*a*) an arbitrarily assumed set of social indicators, or (*b*) an identification of the major social, economic and environmental problems for which the decision-maker is responsible.

Attempts to improve the quality of life will ultimately need to deal broadly with elements such as social indicators. But until such a system has been established and relevant data are available, priority on urgent problems based upon a perception of need will provide a workable basis upon which to proceed.

Performance of Functions (Subsystems)

No political issue is more difficult to resolve than is the question of allocating resources so as to achieve immediate gains and yet not to forget the long term needs that can be delayed but not eliminated.

Many activities are means to some ends. Thus transportation, communication, and energy are service functions whose benefits are dependent on the use to which the service is applied. Other activities have a much more direct effect. For example, hospitals, ambulances, and medical doctors and nurses can have a direct effect on longevity—an important element in the quality of life.

For activities that provide the means (e.g., transportation) rather than directly serving a designated end (e.g., medical), the development of long-term needs is a continuing problem. One reason has been the failure to develop the benefits of proposed improvements first in terms of changes in system output, and then in terms of impacts on the characteristics of life qualities. Analytic capability for producing such estimates was limited in the early 1970s and consisted primarily of empirical data or simply technical judgment.

Social improvements, as noted earlier, can come only through changes in functional (subsystem) *outputs*. Changes in hardware systems, increases in input, or increases in efficiency are important only in the way that changes are produced in the output, or in a reduction of input (thereby, releasing resources).

Estimating the change in a subsystem output due to a change in input (or in efficiency) is a part of the technical process and cannot be produced by political judgment. The classification of the output of subsystems is influenced by the characteristization and purposes of the subsystem. In general, however, the following major categories are descriptive of most subsystems:

- Cost (to the user of the subsystem).
- Time consumed or saved (including reliability).
- Safety (or health-related characteristics).
- User satisfaction (or the user reaction to the quality of service of the subsystem activity).
- Indirect effects (or the external effects that occur as the result of but not related to the purposes of the subsystem).

Given comprehensive data on subsystem performance, and given the results of the feasibility analyses (Chapter 3), the problems of short-term versus long-term benefits can be more effectively solved.

DISTRIBUTION AND DISSEMINATION OF POLICY

The manner in which power is exercised varies with the political structure. But even in the most authoritarian society, political power is limited, in part, by the size of the population over which complete control can be wielded. As the group grows from the size of a family to a national community involving hundreds of millions of people, the power of an individual or a political administration is much less than that necessary to overcome individual resistance, or more aptly, individual decisions.

A national government can control the use of significant levels and percentages of resources. In the United States in the early 1970s, the federal budget was over 20 per cent of the Gross National Product (GNP). Additionally, through laws and regulation, large blocks of resources were controlled or influenced. For example, by Executive action, the federal safety standards were set for highway vehicles that produced an allocation of private sector resources of over $10 billion per year to highway transportation—or approximately 1 per cent of the GNP.

Influences in private sector are more complex, because of the overlapping and diverse interests and desires of individuals. Usually, policies of private companies are dominated by short-term needs to remain financially solvent. The capability of a private company (or individual) to provide nonmaterial benefits in a free enterprise economy is all but impossible. If an item of goods or service is made safer, more esthetic, more durable, or less disruptive of the environment, a private company will probably not be able to survive. Purchasers are either unwilling or unable to distinguish with reliability whether inefficient procedures, high profits, or better products and service are responsible for the difference in costs. Coalitions between major manufacturers are outlawed because of the possibility of collusion, price-fixing, or other monopolistic interferences with a market-place economy.

To offset the limitations of the free enterprise system, governments intercede in the public interest. For societies as large and complex as the United States, and with technology producing increasingly complicated hardware, government has assumed more and more responsibility for protection of the individual and of the public interest. This protection, of course, has its price. What is a protection for some is simply a restriction for others. In this sense, no group can achieve the idealistic concept that each individual will pursue his own destiny. Controls are necessary in the hope of more nearly achieving an ideal life for many.

Certainly for problems of national consequence, the major political power is within the federal government. The initiation of major programs; the regulations of individuals, industries and local government; and the allocation of federal funds to specific projects provide significant bases for controlling the quality of living of individuals.

Because of the power of the federal government, other levels of government and the private sector will need guidelines whenever (*a*) solutions to problems are too great to achieve with the political power available, and/or (*b*) other local governments or industries are influenced by the decision. An individual company will not be able to survive if it acts alone to provide for safer, more durable products or to produce less pollutants. For local governments that are subject to the use of products manufactured and transported across the boundaries of their jurisdiction, assistance from a higher level of government will be needed.

But within each individual, each company and each level of government, some political power exists to improve life qualities. While it is inevitable that outside assistance is helpful in setting general limits or standards, an excessive amount of outside assistance will lead to harmful interference with private and local interests. Particularly the setting of priorities for the solution of problems will be most effective when evaluated at or near the community involved. While national priorities are needed for major elements, local priorities can and do account for most of the local needs.

Figure 9.8 illustrates the problem of obtaining decisions (at any level) that are responsive to overall, societal system requirements. Typical of societal concerns (regardless of the level of the decision) are pollution and

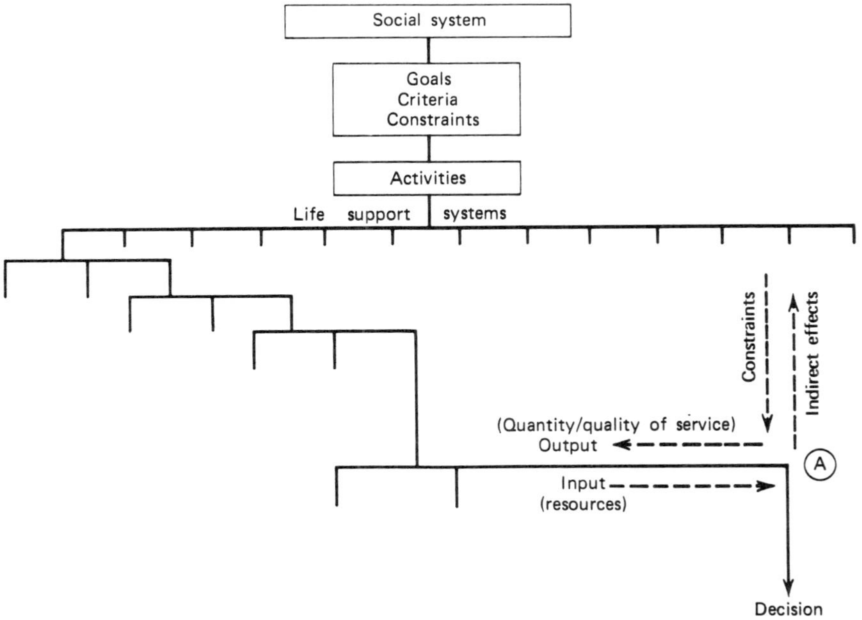

Figure 9.8. Policy distribution.

safety. Obviously, constraints will influence the cost and direct effects (and benefits) from any activity. If the constraints are inadvertently or unnecessarily severe, costs will increase and some other benefits will be sacrificed.

To cope with this level of political issue, it is not practical or desirable to simply extend the normal political process to the smallest levels of the social system. But to avoid a misuse of resources and to increase the effectiveness of policy decisions, the political system should establish workable guidelines. This can be accomplished, in part, by political control of changes in system output (rather than system operation), and in part by setting the constraints within which a system must function.

Referring again to personal safety as an example, the acceptable level of risk (see Chapter 4) for a society could be established by a policy decision. With such a policy, questions in the public and private sectors relative to safety could be resolved in a consistent manner. Risks from hazardous products, working conditions, vehicles, recreation, and such, could be resolved. Thereafter, where the public interest was involved, the question of how safe is safe enough (a political issue) would be defined quantitatively in terms of risk and not be a function of the individual political views of layman or technologists.

Figure 9.8 also illustrates the problem created by undue constraints on subsystems. Thus constraints such as pollution controls, product safety requirements, and such, will influence the direct effects of a subsystem (quantity and quality of service) that contribute to the social system through a network of other subsystems. Unless the interactions produced by the direct effects are understood, limitations on subsystem effects will lead to unexpected and undesired impacts.

Also illustrated in Figure 9.8 is the problem of estimating impacts on the social system. If only the indirect effects are considered, the social impact is less difficult to estimate. But the total impacts of the subsystem reach the social system through a different network. In Chapter 5, the example of policy development for use of underground space dealt with this problem by considering eight life-support subsystems to differentiate between the ultimate impacts of various activities.

With additional studies, and more effective policy decisions, the same type of controls can be applied as needed or desired. While this will not (nor should not) eliminate political debate, it will not require inconsistent and inadvertent political judgment.

The foregoing perspective is foreign to the type of activity wherein each individual attempts to represent the "public interest," and in so doing allocate some level of resources in accordance with his individual preferences. To the extent that these preferences are in conflict with social policy they are counterproductive, and undermine the effectiveness of the political system.

The distribution of policy requirements is no more a usurpation of individual prerogatives than is any other government action. Failure to arrange for effective distribution results, in effect, in political judgements throughout the society. If the society can tolerate these individual and quite variable judgements, then less control by government is desirable. But even then, political judgements should not be marked as technological imperatives.

In the midtwentieth century, many such private, political judgements were made and debated by technologists as technical issues. Abortion, safety, and pollution were particularly sensitive to such treatment. While each of these issues have major technical components, in the final analysis, the problems are resolved on political not technical grounds.

Technical experts who wish to act as advocates for or against a policy issue will need to be aware of how the political system operates. As an advocate, his field of technical expertise will have little to offer. The ground rules of the political process will need to be recognized. It should be neither wrong nor right for a professional or anyone else to be an advocate. But most importantly, his convictions for one side of a political issue are in part related to his own ranking of social priorities or elements of the quality of life.

In this area, needless to say, a technical expert has no special capability by virtue of education or experience. Equally critical is the use of a professional position in a way that undermines the community goals. Such action is not a technical but a social failure. In regard to the field of medicine, for example, doctors cannot practice mercy killings without being in conflict with laws governing such practice. Thus there is no scientific basis for establishing society rules.

The preceding problems of policy distribution and the limitations that policy imposes on technical judgements merely states the basis on which societies will need to operate to avoid what was described in Chapter 1 as the tyranny of technology.

THE FUTURE OF POLICY DEVELOPMENT

Policy development, defined as the analyses upon which resource allocation is based, has existed for centuries and will obviously continue into the future. Policy analyses that are quite relevant to resource allocation have increased both in sophistication and in scope.

In the early part of the twentieth century, the development of modern economic theory provided a quantum improvement. In the midtwentieth century, environmental impact, technology forecasting, and technology assessment emerged as tools for the policy-maker. In many cases, these analyses were directed to the problem of resource allocation. In other instances, the objective was to provide a basis for limiting or constraining the development and use of technology that was deemed to be in conflict with the public interest. While even this latter type of policy analysis was not directed specifically to the problem of resource allocation, the results nonetheless could be used by policy-makers in a way that would have and did have a major influence on the Nation's use of its resources in the 1960 to 1970 decade.

The focus of this book has been on policy development to emphasize, and to identify quite clearly, the process that precedes resource allocation. That such a process must exist is self-evident. That such a process could be improved should not be too surprising. That scientists, technologists, and engineers are needed in the process may at first be surprising. But the idea is not really new, of course. In 1932, the theologian-philosopher, Reinhold Niebuhr wrote: "The realm of politics is a twilight zone where ethical and technical issues meet."[5]

And so we have tried to illuminate this twilight zone—the linking of the technical and political processes is an essential key to the problem of how to use the resources of a society in the most effective way. How "effective way" is defined is irrelevant—whatever path a society chooses to take will be less tortuous, less distracted, and more direct if the linkage is made.

SELECTED READINGS

9.1 *Alternate Federal Policies Affecting the Use of Technology*, Massachusetts Institute of Technology, December, 1972 (PB-221-994).

9.2 Baker, R. F. and R. M. Wilmotte, *Technology and Decisions in Airport Access*, Special Report, American Society of Civil Engineers, 1970.

9.3 Chen, J. (Ed.), *National Priorities*, San Francisco Press, 1970.

9.4 "Goals of Civil Engineering Research—Its Responsiveness to the Needs, Desires and Aspirations of Man," report by Committee on Research, *Engineering Issues*, American Society of Civil Engineers, Vol. 98, No. PP4, 1972.

9.5 Purdee, F. S., et al., *Measurement and Evaluation of Transportation System Effectiveness*, Santa Monica, Calif.: The Rand Corporation, 1969 (PB-185 711).

9.6 *Research Needs in Civil Engineering Relevant to the Goals of Society*, Special Report, American Society of Civil Engineers, 1971.

[5] Reinhold Niebuhr, *Moral Man in an Immoral Society*, New York: Charles Scribner's Sons, 1932.

9.7 *Social Goals and Indicators for American Society*, The Annals of the American Academy of Political and Social Sciences, Vol. I and II, Vol. 371 and 373, May and September, 1967.

9.8 *Social Indicators—1973*, Office of Management and Budget, Executive Office of the President, U.S. Government Printing Office, 1973.

9.9 *Toward Balanced Growth: Quantity with Quality*, Report of the National Goals Research Staff, Executive Office of the President, 1970.

INDEX